Auditing in the food industry

Related titles from Woodhead's food science, technology and nutrition list:

Making the most of HACCP: Learning from others' experience (ISBN: 1 85573 504 0)

Written by those who have actually implemented HACCP systems, this authoritative collection reviews the opportunities and pitfalls in setting up and running HACCP systems in practice. An essential guide both for those implementing HACCP systems and those who want to know how to improve the systems they have in place.

Food process modelling (ISBN: 1 85573 565 2)

A major trend in the food industry has been the use of modelling techniques to measure, predict and control food processes in the search for improved safety and quality. *Food process modelling* provides an authoritative review both of key modelling principles and its practical application in such areas as microbiological safety, raw material production, processing, storage and distribution.

Instrumentation and sensors for the food industry Second edition (ISBN: 1 85573 560 1)

The first edition of this book established itself as a standard review of instrumentation for the assessment of food safety and quality. The second edition consolidates this reputation, taking in new areas such as biosensors.

Details of these books and a complete list of Woodhead's food science, technology and nutrition titles can be obtained by:

- visiting our web site at www.woodhead-publishing.com
- contacting Customer services (email: sales@woodhead-publishing.com; fax: +44 (0) 1223 893694; tel.: +44 (0) 1223 891358 ext. 30; address: Woodhead Publishing Limited, Abington Hall, Abington, Cambridge CB1 6AH, England)

If you would like to receive information on forthcoming titles in this area, please send your address details to: Francis Dodds (address, tel. and fax as above; e-mail: francisd@woodhead-publishing.com). Please confirm which subject areas you are interested in.

Auditing in the food industry

From safety and quality to environmental and other audits

Edited by
Mike Dillon and Chris Griffith

CRC Press
Boca Raton Boston New York Washington, DC

WOODHEAD PUBLISHING LIMITED
Cambridge England

Published by Woodhead Publishing Limited
Abington Hall, Abington
Cambridge CB1 6AH
England
www.woodhead-publishing.com

Published in North and South America by CRC Press LLC
2000 Corporate Blvd, NW
Boca Raton FL 33431
USA

First published 2001, Woodhead Publishing Limited and CRC Press LLC

British Library Cataloguing in Publication Data
A catalogue record for this book is available from the British Library.

Library of Congress Cataloging-in-Publication Data
A catalog record for this book is available from the Library of Congress.

Woodhead Publishing Limited ISBN 1 85573 450 8
CRC Press ISBN 0-8493-1214-0
CRC Press order number: WP1214

Cover design by The ColourStudio
Project managed by Macfarlane Production Services, Markyate, Hertfordshire
(email: macfarl@aol.com)
Typeset by MHL Typesetting Limited, Coventry, Warwickshire
Printed by TJ International, Padstow, Cornwall, England

Contents

Contributors

Chapters 1 and 2

Dr Mike Dillon
Mike Dillon Associates Ltd
32a Hainton Avenue
Grimsby DN32 9BB
England

Tel/Fax: +44 (0)1472 348852
E-mail: *mdillon@mdassoc.demon.co.uk*

Chapter 3

Ms Sue Dix
Tesco Stores plc
PO Box 400
Cirrus Building
Shire Park
Welwyn Garden City AL7 1AB
England

Tel: +44 (0)1707 634932

Chapter 4

Vance McEachern, Alfred Bungay, Shelley Bray Ippolito and Sue Lee-Spiegelberg
Canadian Food Inspection Agency
59 Camelot Drive
Nepean
Ontario K1A 0Y9
Canada

Tel: (1) 613 225 2342
Fax: (1) 613 228 6654
E-mail: *VmcEachern@EM.AGR.CA; SFLEE@EM.AGR.CA; ABUNGAY@EM.AGR.CA*

Chapter 5

Mr Malcolm Kane
The Copse
30 Brewery Road
Pampisford
Cambridge CB2 4EN
England

Fax: +44 (0)1223 830918

Chapter 6

Dr David Rose
Campden and Chorleywood Food Research Association
Chipping Campden GL55 6LD
England

Tel: +44 (0)1386 842088
Fax: +44 (0)1386 842100
E-mail: *d.rose@campden.co.uk*

Chapter 7

Dr Neil Khandke
Ice Cream Technology Unit
Unilever Research
Colworth House
Sharnbrook
Bedford MK44 1LQ
England

Fax: +44 (0)1234 222000
E-mail: *Neil.Khandke@unilever.com*

Chapter 8

Dr Roger Wood
Food Standards Agency
c/o Institute of Food Research
Norwich Research Park
Colney
Norwich NR4 7UA
England

Tel: +44 (0)1603 255298
Fax: +44 (0)1603 507723
E-mail:
roger.wood@foodstandards.gsi.gov.uk

Chapter 9

Dr Dotun Adebanjo
Leatherhead Food Research Association
Randalls Road
Leatherhead KT22 7RY
England

Fax: +44 (0)1372 386228

Chapter 10

Mr Par Olsson
The Swedish Institute for Food and Biotechnology
PO Box 5401
SE-402 29 Göteborg
Sweden

Fax: +46 31 83 37 82
E-mail: *par.olsson@sik.se*

Chapter 11

Mr John Parslow
The Soil Association
Bristol House
40–56 Victoria Street
Bristol BS1 6BY
England

Fax: +44 (0)117 925 2504
E-mail: *j.parslow@soilassociation.org*

1

Introduction

The food industry faces an unprecedented level of scrutiny. Consumer concerns about safety have prompted an increasing level of regulation. Customers have ever higher expectations of quality, and food manufacturers have responded by developing systems to measure, manage and improve product quality more effectively. At the same time, there has been a shift in the relationship between the industry and those setting and enforcing standards. The traditional approach was a paternalistic one, with regulators setting and enforcing prescriptive standards, and food manufacturers responding retrospectively to infringements identified by regulatory inspections. This approach has been increasingly seen as inflexible and reactive. It is being replaced by a new relationship in which, within a framework of appropriate standards, food manufacturers take greater responsibility for the safety and quality of their products.

The role of the regulator in this new relationship moves away from inspection of specific techniques and products to auditing the system set up by businesses to manage safety or quality proactively to prevent problems. With this shift, auditing has become a key issue for the food industry, from how regulators audit food businesses effectively, to how food businesses audit themselves to improve their performance.

One of the main factors behind the increase in auditing experienced by food manufacturers is the demand of retailers. Retailers want to know that the products they sell are of a specified and consistent quality and are safe. This increased level of auditing poses problems for both retailers and those manufacturers supplying them. A plant producing products for more than one retailer might find itself being audited repeatedly by auditors from differing retailers, disrupting its operations and facing potentially conflicting recommendations from differing auditors. With their huge international supplier base, retailers face the prospect of having to employ a large auditing team. There have been two recent developments to resolve these problems:

1. the expansion in third-party accreditation
2. the British Retail Consortium (BRC) Technical Standard.

The use of accredited third-party auditors reduces the likely overlap of effort by separate retailer audit teams. It also addresses potential concerns about the impartiality, consistency and quality of audit, providing there is an adequate accreditation framework. In the UK, the United Kingdom Accreditation Service (UKAS) is responsible for the accreditation of laboratories undertaking analysis for quality control, auditors and certification bodies across a broad range of activities. UKAS can evaluate the competence and impartiality of auditors against international standards such as EN45004. UKAS approval allows approved organisations to carry an accreditation logo. In turn, UKAS is recognised by international agreements, reducing the need for auditing and inspection in other countries. In Europe, for example, the status of UKAS as a national accreditation body is recognised by its membership of European Cooperation for Accreditation (EA). It has been suggested in some countries that government auditing activities, for example inspections of food processing facilities currently undertaken by local Environmental Health Officers (EHOs) in the UK, could use the same model.

If the use of third-party auditors addresses the question of who might undertake an audit most effectively, the BRC Technical Standard addresses that of establishing a common standard amongst retailers. In the UK such a standard is particularly significant as the food industry is dominated by a small group of large retailers, with over 50% of all food sold being retailer branded products. Prior to the BRC Technical Standard, each retailer audited businesses against its own internal standards, resulting in inconsistencies and occasionally conflicting audit findings and recommendations from different retailers auditing the same business. Audits of compliance with the standard must be by auditors who comply with EN45004 and are accredited by a national accreditation body such as UKAS.

This collection explores the various dimensions of auditing in the food industry. A first chapter sets the scene by exploring the range of standards in the food industry and introducing the principles and practices of auditing. This chapter leads on to Chapters 3 and 4 which look at auditing from the auditor's perspective, firstly from the point of view of retailers auditing their suppliers, and secondly from government auditing of the food industry. Part II looks specifically at safety and quality. There are chapters on how retailers and manufacturers audit HACCP systems, the auditing of TQM systems and quality control of microbiological analysis, an essential part of any safety or quality system. Finally, Part III looks at newer types of audit that are gaining in importance within the food industry, ranging from benchmarking to environmental and organic audits. It is hoped that the book will help strengthen existing auditing skills and develop skills in new areas to the benefit of consumers, government and industry alike.

Part I

The auditing process

2

Food standards and auditing

M. Dillon, Mike Dillon Associates Limited, Grimsby

2.1 Introduction: why have standards become so important?

This book sets out to review the necessary systematic and detailed approaches to verifying that standards have been consistently met through an audit process. Different types of standard and approaches to their audit are discussed in relation to the food sector. This first chapter provides an insight for the reader into the types of standards, the principles of their development and the important role of the Codex group.

The increased need to meet a wide range of standards by trade partners to ensure 'equivalence' in food control or achieve market expectations has elevated the commercial importance of food standards. The rising actual and perceived economic impact of meeting these standards has necessitated investigations into the relevant costs and benefits.[1,2] Many developing countries have claimed standards are in effect trade barriers, as the countries imposing such standards often do not comply with rigorous controls internally, e.g. zero tolerance for Listeria within meat products in the USA.[3] The meat lobby within the USA is claimed to demand such standards to prevent the import of competitor products. The increasing number of widely reported serious food poisoning outbreaks has also raised public demands for more effective food standards.[4,5]

Transparency within the standard setting and compliance process is therefore crucial to international trade. Emberley has reported that there is a need for the 'fair validation' of food control systems to prevent non-tariff trade barriers and ensure their equivalency.[6] This has resulted in an increasing workload for the Codex Committee on Food Import and Export Inspection and Certification (CCFICS). This committee developed and reviewed the criteria for inter-country inspections and has thus clarified the mechanisms for mutual recognition agreements (MRAs) and memorandums of understanding (MOUs) as the basis of bilateral food trade. The mutual recognition of equivalent standards and conformity assessment procedures has moulded international standards guidelines and requirements in the last five years.

Audit systems are now used by government agencies who monitor these agreements and therefore audit has been recognised as the mechanism to ensure effective food control systems have been implemented and maintained. The United States has prioritised the development of MOUs with its major trading partners particularly those with established operational hazard analysis critical control point (HACCP) systems. The route to determining equivalence is now based on desk and on-site audit techniques. Apart from this legal drive for audit, suppliers also want to meet agreed market standards and therefore face a larger number of 'supplier' audits from their customer base.

2.2 What are standards?

Standards are agreed sets of criteria for ensuring consistent manufacture of food products from a safety, nutritional or management system perspective. These standards may be required by law or by the market. There are a wide range of legal food standards, ranging from product specific such as meat, fish and eggs (EU directives) to cross cutting, such as the horizontal general hygiene directive. These standards are therefore established by agreement and approved by a recognised body that ensures consistent manufacture within agreed rules. Completed standards should be simply documented technical specifications aimed at the promotion of community benefits.[7]

2.2.1 Why have standards?

In complex food networks clear standards are needed to ensure efficient food production within a myriad of legal requirements. The *ISO/IEC Guide* reported the improved suitability of the product, process or service for an intended purpose when standards are used. Furthermore, standards should prevent barriers to trade and enable technological co-operation. Standards may therefore be focused on, but not restricted to, variety control and protection of the product, consumer or environment, and also cover health and safety.

Some customers may only require a 'certificate of conformity' to guarantee that an agreed specification is being met. Many companies now insist on a 'supplier appraisal' scheme to compare performance against agreed quality and legal standards. Much of the world food trade is reliant on raw material from developing countries.

The International Trade Centre (ITC) is the focal point in the United Nations system for technical co-operation with developing countries in trade promotion. The ITC was created by the General Agreement on Tariffs and Trade (GATT) in 1964, and since 1968 has been operated jointly by GATT and the UN. A specific focus of the ITC is to enhance the ability of target countries to penetrate the international export market. This has involved providing guidance to governments and small to medium-sized enterprises (SMEs) on achieving specific rigorous food standards and creating equivalent national standards.

2.2.2 Principles outlined in the standard setting process

Simplification

This is a continual process of converting complex processes into an easily understood model, e.g. the Institute of Food Science and Technology (IFST) *Good Manufacturing Practice* guidelines.[8] This document provides guidance to food manufacturers on how to manufacture safe, quality food products consistently by adherence to the principles of GMP within a wide range of processes.

Co-operation

The standard should be reached by consensus between the relevant stakeholders, as economic and social requirements may be in conflict. The process by which standards are agreed has been used to incorporate a risk based approach, that is understood and accepted by the population. The co-operation and involvement of the public in agreeing acceptable levels of risk within a given food standard is now an essential part of setting standards and avoiding conflict.[9]

Implementation

The standard must be achievable and clear guidance on implementation provided. The original World Health Organisation (WHO) documents on food safety systems, i.e. HACCP, contained principles, but insufficient information was provided on how to achieve them consistently.[10] More recent publications have provided guidance on the methods and approaches required to achieve standards.[11, 12]

Selection

The areas and subjects that standards may cover in the food sector are wide and must be chosen carefully. The areas of focus and the basis of the standard are selected specifically to ensure that topical issues are encompassed in any new development and the standard will remain relevant for as long as possible without revision.

Revision

Normally, standards are reviewed at regular intervals, usually every three years. Currently, a major review of the food safety hygiene rules is being undertaken by the European Union. Under the proposals announced in Brussels on 17 July 2000, food operators throughout the chain will bear primary responsibility for food safety. The proposals are contained within four new regulations, which will merge, harmonise and simplify the complex hygiene requirements contained within the 17 previous directives. The standards are becoming goal orientated, i.e. objectives are set which businesses may flexibly meet. The regulation, once adopted by the European Parliament and Council, will replace the 16 product specific directives and 93/43/EEC, which is the horizontal directive on the hygiene of foodstuffs.

This revision introduces the farm to fork principle within hygiene policy, and includes programmed self-checking and modern hazard control techniques. Food producers assume primary responsibility for food safety. The implementation of harmonised HACCP systems by non-primary food operators will also become a mandatory requirement. A third principle will be traceability of all food and food ingredients. These new requirements will result in the fourth principle, which is the compulsory registration of all food businesses. These revisions have been made to keep the existing standards relevant and capable of achieving consistent control within the food chain.[13]

Determination of compliance

When the key process or product criterion is defined, the specifications must include a description of the recommended or compulsory methods and tests. Sample size, frequency and method should be given if necessary. Existing EU directives and annexes often give guidance on specific methods to be used to determine compliance (European Commission Council Directive 80/778/EEC).

This difficulty in assessing compliance should be considered when confirmation of control is necessary within a food processing plant holding 3,000 ingredients, which may end up in 300 different products. Determination of compliance with labelling, food safety and environmental standards within this operation will require dedicated management control and monitoring. The new legislation on genetically modified organisms (GMOs) may require an understanding not only of the raw materials, but also of the medium on which they are produced, e.g. xanthan gum may be produced on soya medium. The traceability principle therefore becomes important in ensuring no unacceptable contaminants, possible allergens or genetically modified material are present. A recent dioxin case involved 8 kg of material, which was moved to more than 400 products and the overall incident was estimated to cost $2 billion dollars.[14]

Legal enforcement

This is dependent on society and the national opinion on the need for legally enforceable standards. In many cases, countries vacillate between encouraging members of the food chain to comply with relevant standards and strict enforcement of standards.

2.3 Standards and specifications

The term ‘standard’ can be used in different ways. It can refer to a relative level, i.e. the standard of something, or to an absolute threshold, e.g. minimum performance criteria. Food standards are issued by government bodies, and companies must comply with them. They safeguard public health, safety and the environment and are therefore compulsory. United Kingdom food legislation requires food to be not injurious to health. Some US and other legislation requires food to achieve specific microbial standards.[15]

2.3.1 Mandatory standards

Examples of mandatory standards include the absence of Salmonella, Cholera and other pathogens in processed foods and maximum residue for chemicals, heavy metals and specific pesticides. These standards specify limits for toxicants, pathogens and additives. Additional standards cover packaging and labelling.

2.3.2 Microbiological process standards

Garrett has reported a shift away from 'end product microbiological testing to determining individual lot compliance – towards microbiological in plant process standards'.[9] The author explains that this was not a new concept in the USA where defect action limits (DALs) were proposed by the Food and Drug Administration (FDA). This required specific sampling at identified points during the process with the sub-samples analysed for APC, E.coli, and Staphylococcus aureus. The United States Department of Agriculture (USDA) are currently operating a pathogen reduction programme which requires in-plant microbiological Salmonella performance standards for 'steers/heifers, cows/bulls, ground beef, hogs, broilers, ground chicken, and ground turkey' relative to Salmonella. The tolerances required for Salmonella within the sampling plan are based on baseline data collection and surveys developed by the Food Safety Inspection Service (FSIS). A key reference benchmark was then available for the agency to measure the impact of their regulatory HACCP based programme. Garrett therefore believes that zero tolerance for specific pathogens may be replaced by regulatory performance standards.

2.3.3 Transparency in Food Standards

Garrett then reviews the concept of 'transparent', which was defined by a working group composed of International Commission on Microbiological Safety for Foods (ICMFS), National Advisory Committee on Microbiological Criteria for Food (NACMF) and Codex.[16] Mitchell has also stressed the need for microbiological risk analysis in his recent publication.[17]

Transparency is defined by Garrett as 'Characteristics of a process where the rationale, the logic of development, constraints, assumptions, value judgements, decisions, limitations and uncertainties of measurement are fully and systematically stated, documented and accessible for review'. This is reflected in the increasing focus within audit on the 'hazard analysis' component of key food safety systems. The Campden & Chorleywood Food Research Association (CCFRA) document in the UK is an attempt to inform industry of how this could be approached in a systematic manner. International trade will also increasingly rely on audit to ensure that equivalency and transparency are 'standardised'.

2.3.4 Voluntary standards

These standards are set by trade associations or companies and detail minimum requirements for products, but are not legally enforceable. They are intended to provide a guide for business to achieve target requirements, often in relation to market needs.

Extracts from the United Nations guide to GMP for fish processors are provided in the Appendix to this chapter.[18] These provide a more detailed review of voluntary standards. The USA Seafood Alliance also operated a voluntary standard for HACCP within the seafood sector prior to the mandatory standard defined in law in 1997. This voluntary standard also includes a detailed approach to a pathogen reduction programme, which has now been included in the mandatory requirements of classes of American food standards.[19]

Standards are viewed as the minimum requirement, but specifications are often much more stringent. Often problems may arise when enforcement officers inspect non-compliance records, which demonstrate that the business may not be meeting current market requirements, but achieves the minimum legal requirements. Companies may hold specifications for raw materials, in process and final product specifications.

Companies must meet an increased range of market specific standards. The UK retailer requirements for food control are outlined in the British Retail Consortium (BRC) standard.[20] The ISO series may be a mandatory requirement for supply or operation of a specific aspect of the supply chain, e.g. ISO 14000, 9000 (2000). Alternatively, environmental standards have been incorporated into these market specific requirements, e.g. Tesco's Nature's Choice encompasses environmental requirements.

The reader will be directed to further sources of information on these standards, as the objective here is to place the standards in context prior to descriptions of the necessary audit process.

2.3.5 The British Retail Consortium (BRC) Technical Standard

This standard is an example of a widely used voluntary standard in the UK. It has been developed by the British Retail Consortium, which represents the major retailers in the UK, as a common standard for manufacturers supplying goods to BRC members. The standard is structured into six broad areas covering HACCP, quality management systems, factory environmental standards, product and process control, and personnel. Each section begins with a statement of intent, for example for HACCP:

The basis of the company's food safety control system shall be a HACCP plan which shall be systematic, comprehensive and thorough and shall be based on the Codex Alimentarius HACCP principles.

All suppliers must comply with the statement of intent in order to gain a certificate of inspection. The standard sets out two levels, Foundation and Higher, as well as recommendations for good practice. It is intended that companies progress from one level to the other and, ultimately, comply with all guidelines. The standard sets out the frequency of inspection and requires that inspection reports record non-conformities against all three columns. Auditing of compliance with the standard must be by inspectors who comply with EN45004 and are accredited by a recognised national accreditation body, which is UKAS in the UK. The standard requires that suppliers address all non-conformances, take appropriate corrective actions and maintain on-going surveillance of the effectiveness of their safety and quality systems.

The standard has a number of advantages:

- it provides a single standard and protocol for retailers and suppliers to follow
- it is comprehensive in scope, covering product safety and quality as well as environmental issues
- it covers the scope and frequency of inspection, and the standard required by auditors.

2.4 Increasing importance of HACCP based Codex standards (GATT)

The importance of standards for the movement of products in the international arena was heightened after the 1994 GATT agreement. The decision by the World Trade Organisation (WTO) to require a risk based approach to standard setting indirectly emphasised the importance of the HACCP based approach to food control. The general guidance provided by Codex on HACCP in 1993 was required to be placed within product specific standards such as meat and fish.

The Codex body has been described by Mitchell as the grandparent of standard setting organisations.[21] The structure provided by Codex, within the general hygiene directive, provides the minimum requirements for control within many supporting prerequisite programmes such as pest control. The guidelines provided by Codex have largely been adopted by some developing countries in the past as the basis for their own legislation. Additionally, the Codex group standards are used by the extension arm of the United Nations, the Food and Agriculture Organisation (FAO), as the basis for training and technical co-operation interventions designed to assist countries to meet given standards.[22]

2.4.1 International standard setting body – Codex Alimentarius Commission

The Codex Alimentarius Commission (CAC) was set up in 1962 by the WHO and the FAO to implement the joint FAO/WHO Food Standards Programme. The 1994 GATT agreement has made this group one of the most influential in the development of food standards governing international trade. The twin purpose of the programme is to protect public health and ensure fair practices in trade through development of standards that may then be modified or used by international government or non-government bodies.

2.4.2 GATT – Uruguay round final act, Marraksah 1994

The final act in the Uruguay round of GATT involved the agreement and application of sanitary and phytosanitary measures (SPS) on technical barriers to trade. The SPS agreement placed Codex standards, guidelines and recommendations as the 'specifically identified baseline' for consumer protection under this agreement. GATT member countries may introduce higher level SPS protection if there is a 'scientific justification'. The WTO acts as a regulatory body to settle international trade disputes over SPS issues. The WTO has the power to fine or impose trade penalties on GATT countries that do not comply with the SPS. Mitchell reported in 1996 that this would necessitate specific issues to be investigated and negotiated in the context of HACCP.

2.4.3 CAC Food Hygiene Committee

An *ad hoc* international working group produced a common definition of the principles of HACCP and their application to food operations in 1993. This has now been recognised as a landmark document and was initially believed to facilitate trade at national and international level. The document outlined the seven principles of HACCP and discussed a 12-stage process for effective adoption. The principles were accompanied by a logic tree to assist the user in the consistent identification of critical control points. The Food Hygiene Committee has since reviewed and reformulated the Codex general principles of food hygiene, which is the bedrock of the other food hygiene codes. The October 2000 meeting of the CAC committee again re-focused the HACCP guidance document to ensure its relevance to smaller food organisations.[23]

2.5 European Union standards

Jukes reviewed the legal structure of the Community (as it was then called) and emphasised the importance of the Treaty of Rome as the basis of Community activity.[24] It is worth noting that although the major emphasis of articles within the treaty is to enable free market access and encourage trade, Article 36 allows member states to prohibit imports on grounds of public health. The Single European Act, which proposed to encourage closer co-operation within the Community has also been viewed as a mechanism for strengthening external border control.[25, 26]

The adoption of increasingly stringent and enforced food standards at points of entry into the EU has resulted in the closure of the market to many developing countries. The audit process employed by the EU to evaluate the equivalence of competent authorities in trade partners has also been criticised for being non-transparent. Lima Dos Santos and Lupin reported the closure of the market to Bangladeshi shrimp caused by the inability of the Bangladeshi government and industry to meet the standards demanded by the European Union under 91/493 (Fish Specific Directive).[27]

2.5.1 Single administrative document (SAD)

The distribution of goods within the European Union is accompanied by a single administrative document, which promotes a harmonised system of plant health and veterinary inspection. Depending upon the foodstuffs entering the market, the port health authority will inspect cargoes, normally to ensure that a 'certificate of conformance' to agreed specifications is in existence. For products with a known problem history, e.g. desiccated coconut, further samples may be required.

2.5.2 Union legislation

Union legislation covering foodstuffs consists of regulations and directives. EU regulations have the force of law in all member states, have a legal binding effect, prevailing over national legislation, and do not have to be ratified by national parliaments. Directives have no direct force of law and must be translated into existing national law. This is achieved by the Food Safety Act in the UK where these directives are translated into statutory instruments.

2.6 UK Food Safety Act

The UK Food Safety Act 1990 was designed to consolidate, modernise and strengthen existing food legislation, and concentrated on the two major themes of the protection of the consumer from fraud or adulteration, and the assurance of public health. The Act enables the rapid implementation of EU directives. The due diligence defence incorporated within Section 21 of the Act places the onus of proof upon the food company to demonstrate that 'all reasonable precautions and all due diligence has been exercised in relation to the offence'.

Blanchfield provided an analysis of the defence.[28] This proposed that 'due diligence' should be regarded as the management activity of ensuring control and that all reasonable precautions would relate to the control measures required by the standard. Dillon provided a full explanation of the defence with examples of failure to meet the agreed standards and resulting fines.[26] Effectively, this legislation has meant that industry is 'guilty until proven innocent', resulting in the raising of standards throughout the chain. Industry has interpreted the legislative standard as necessitating the implementation of a formal quality management system focused through risk analysis provided by HACCP. UK industry, and therefore its international supply chain, has adopted formal quality assurance (QA) standards such as the ISO series accompanied by HACCP systems.

2.7 The need for audit

Industry and governments have realised that effective food control systems require shared responsibility in aspects of their design, operation and verification. Governments are required to set the overarching limits within which these systems operate, and industry must design and operate to meet these limits. Food control standards, once set up, are not always effectively implemented, because of resistance to change, lack of commitment, limited resources and increased training requirements. Commercial advantages have not always been measured or promoted and industry fears the cost implications in achieving set standards. Audit has often been poorly addressed by government and industry. Confusion exists about the meaning of the words audit and verification, both in the US and in other countries.

The role of audit is therefore of increasing importance, and the relevant skill sets required by food 'inspectors' need to be defined and agreed at national and international level. Auditing food control systems using standard methods is now recognised as a challenge for the new millennium, for both industry and government, in the expanding and increasingly complex world of food protection. It is a challenge that must be met![29]

2.8 List of useful websites

Chartered Institute of Environmental Health *http://www.cieh.org.uk*

Codex Alimentarius Commission *http://www.fao.org/waicent/faoinfo/economic/esn/codex/Default.htm*

DH/MAFF Joint Food Safety and Standards Group *http://www.doh.gov.uk/jfssg.htm*

European Chilled Food Federation *http://www.chilledfood.org/ecff.htm*

European Union *http://www.europa.eu.int/*

Institute of Food Research *http://www.ifrn.bbsrc.ac.uk*

LACOTS *http://www.lacots.org.uk*

MAFF Food Safety *http://www.maff.gov.uk/food/foodindx.htm*

US Govt Food Safety Information *http://www.foodsafety.gov/*

2.9 References

1. DILLON, M., HANNAH, S. JAMES, T., and THOMPSON, M. 'Cost of Food Control: Building Profitable Business'. Second NSF International Conference on Food Safety, Savannah, Georgia, USA, 2000.
2. GRIFFITH, C.J., MORTLOCK, M.P., and PETERS, A.C. 'Evaluating the Economic Impact of HACCP Implementation in Welsh Butcher Shops'. Second NSF International Conference on Food Safety, Savannah, Georgia, USA, 2000.
3. Report of the FAO Expert Technical Meeting, *The Use of HACCP Principles in Food Control,* Vancouver, 1994.
4. BELL, C., and KYRIAKIDES, A. *E.coli: A Practical Approach to the Organism and its Control in Foods,* London, Blackie Academic & Professional, 1998.
5. BELL, C., and KYRIAKIDES, A. *Listeria: A Practical Approach to the Organism and its Control in Foods,* London, Blackie Academic & Professional, 1998.
6. EMBERLEY, B.J. 'A Global Approach to Determining Equivalency Between Inspection and Certification Systems'. *Fish Inspection, Quality Control and HACCP.* Washington, USA. Technomic, 1996.
7. *ISO/IEC Guide* 2, 5th Edition, Geneva, 1986.
8. INSTITUTE OF FOOD SCIENCE & TECHNOLOGY (UK), *Food and Drink Good Manufacturing Practice: A Guide to its Responsible Management,* London, Institute of Food Science & Technology (UK), 1998.
9. GARRETT, E. 'Along the Yellow Brick Road toward Microbiological Risk Assessment'. The Canadian Food Inspection Agency's Annual Conference, Vancouver, 2000.
10. CODEX ALIMENTARIUS, 'Guidelines for the Application of the Hazard Analysis Critical Control Point System'. ALINORM 93/131, Appendix II, 1993.
11. MORTIMORE, S. and WALLACE, C. *HACCP: A Practical Approach.* London, Chapman and Hall, 1994.
12. VOYSEY, P. *An Introduction to the Practice of Microbiological Risk Assessment for Food Industry Applications*, Guideline No. 28, Gloucester, Campden & Chorleywood Food Research Association Group, 2000.
13. *Fish Inspector*, No. 48, September 2000.
14. CARTER, J. 'Traceability throughout the Supply Chain, Dioxin Case Study'. *Food Traceability and Crisis Management,* London, 2000.
15. DILLON, M. and GRIFFITH, C. *How to Audit: Verifying Food Control Systems.* Grimsby, MD Associates, 1997.
16. CODEX ALIMENTARIUS COMMISSION *Recommended International Code of Practice on the General Principles of Food Hygiene.* Annex II CAC/RCP 1969, Rev 3 M 99/13A Appendix II, 1997.
17. MITCHELL, R.T. *Practical Microbiological Risk Analysis: How to Assess, Manage and Communicate Microbiological Risks in Foods.* Oxford, Chandos Publishing, 2000.
18. NASINYAMA, G., DILLON, M., TIFFNEY, P. and THOMPSON, M. *Manual of Good Manufacturing Code of Practice (GMP) for Fish Factories*, UNIDO Integrated Country Programme for Uganda, May 2000.

19. SPENCER GARRETT, E. and HUDAK-ROOS, M. 'Developing a HACCP based Inspection System for the Seafood Industry'. *Food Technology,* December 1991.
20. BRITISH RETAIL CONSORTIUM, *Technical Standard and Protocol for Companies Supplying Retailer Branded Food Products*, Issue 2. London, British Retail Consortium, 2000.
21. MITCHELL, R.T. 'HACCP: An International Overview', *HACCP: Knowing Your Enemies and Your Friends,* Cardiff Institute of Higher Education, 1996.
22. FAO, *HACCP Training of Trainers Programme*, Rome, February 1995.
23. JANSEN, J.T. 'Application of HACCP Principles in Small and Less Developed Businesses (SLDBS) with a Focus on Recent Discussions in Codex Alimentarius and during WHO Expert Consultations'. Second NSF International Conference on Food Safety, Savannah, Georgia, USA, 2000.
24. JUKES, D.J. Food Legislation of the UK: A concise guide, 3rd edn, Butterworth Heinemann, 1993.
25. DILLON, M. 'Confrontation or Collaboration? The New EC Food Laws'. *Appropriate Technology,* Vol. 19, No, 1, June 1992.
26. DILLON, M. 'Due Diligence – The Implications of the Food Safety Act (1990) to the Development of Appropriate Quality Systems'. *New Markets in Seafood,* Hull International Fisheries Institute, UK, 1993.
27. LIMA DOS SANTOS, C.A. and LUPIN, H.M. 'FAO's Experience in HACCP Training of Regulatory and Industry Personnel'. *Fish Inspection, Quality Control and HACCP*, Washington, USA, Technomic, 1996.
28. BLANCHFIELD, J.R. 'Due Diligence – Defence or system.' *Food Control* No. 3, 1992.
29. DILLON, M. 'Verifying Food Safety Systems in the 1990s'. *The Role of Government Agencies in Assessing HACCP*, FAO/WHO Consultation, Geneva, 1997.

Appendix: Example taken from GMP Manual

9.1 Cleaning schedules documented

9.1.1 Written, formalised cleaning procedures and schedules must be available for every department within the factory. They must be clear, legible and easy to follow.

Auditors recommendations:

Look at:	Look for:
Cleaning procedures	Clarify and legibility Coverage of all areas Availability

9.1.2 The cleaning schedules must dictate the frequency and method of cleaning and disinfecting agents that are to be used for all plant, equipment and surroundings.

Auditors recommendations:

Look at:	Look for:
Cleaning schedules	Detailed methods Chemicals defined Frequency of cleaning

9.1.3 For companies employing different ethnic groups in significant numbers, the cleaning procedures should also be explained in a language that is understandable to the workers. This may best be achieved or supported through picture diagrams.

Auditors recommendations:

Look at:	Look for:
Cleaning instructions	Understanding by staff Language issues

9.2 Approved food grade detergents in use, e.g. taint risks/phenols

9.2.1 All cleaning and disinfection agents used on site must be 'approved' food grade materials, supplied by a reputable company.

Auditors recommendations:

Look at:	Look for:
Cleaning chemicals and data sheets	Food grade marking Reputable supplier

3

What auditors look for

A retailer's perspective

S. Dix, Tesco Stores plc, Welwyn Garden City

3.1 Introduction

Most retailers have produced guidelines and specific codes of practice which detail the management control systems that they expect to find in their suppliers of retailer branded goods. They also spell out the conditions of the structure and equipment that they believe is necessary to comply with legal requirements and ensure product safety consistently. Auditing is simply a check against these criteria to ensure that the supplier is complying with these codes and guidelines and is maintaining the systems and conditions adequately. The retailer has a responsibility and a commercial need to ensure that products carrying their name are safe and legal and that information provided about the product is legal, decent, honest and truthful. To this end, the audit is usually designed primarily to ensure that the supplier has not breached the retailer's guidelines, thereby providing the retailer with a due diligence defence. However, the opportunity can also be taken to identify the ways in which standards can be improved by tightening controls where necessary, thereby moving both the supplier and retailer forward in improving both quality and service for the customer. In consequence, an audit is frequently planned in advance to ensure that the supplier has all of the appropriate people and documentation available on the day.

There are two types of audit:

1. Routine:
   - approval of a new supplier
   - regular check on an existing supplier to ensure continued compliance with the previously approved quality management system. This is usually described as a 'due diligence' audit and is carried out by the retailer or their agent.

2. Non-routine, for the following reasons:
   - serious interruption of supply
   - variable product quality noticed
   - high customer complaint level
   - adverse microbiological trend reported
   - adverse media reports
   - anonymous tip-off
   - product withdrawal.

3.2 Routine auditing: new suppliers

The auditing of a potential supplier should include a thorough examination of their hazard analysis critical control point (HACCP) and quality management systems both on paper and in practice. There must be a thorough inspection of the factory to assess whether or not the standards of maintenance and cleanliness comply with the retailer's code of practice. Many companies would like to supply major retailers so it is common practice to send out a pre-audit questionnaire. The format of this may vary depending upon the type of product manufactured. Risk assessment is often used to define the conditions required, the frequency of re-audit and the number and frequency of product checks that should be carried out by both the supplier and retailer. The pre-audit questionnaire will therefore be highly detailed for suppliers of products such as chilled ready meals which must be manufactured in a high care environment and less detailed for low risk producers such as produce packers. The replies to these questionnaires may eliminate some potential suppliers. However, if there is a real customer need then retailers may send the company a copy of their own quality management systems manual and help them to work through and comply with the contents. This is most likely to occur with a new business unused to dealing with major retailers.

3.2.1 Auditing a supplier's safety and quality systems

One indicator of the suitability of a supplier would be the existence of a fully documented safety and quality management system. This should be detailed in a manual which contains a description of all the control measures taken within the factory and stipulates the frequency with which all checks are carried out, together with the name of the most senior manager accountable for these measures. Systems change continuously within a factory as a response to the introduction of new equipment, ingredients and processes. A retail auditor would therefore expect to see evidence that the management system was under continuous review. This is easily checked by looking at an index showing revision numbers and dates.

Retailers and their agents will usually send a checklist to potential suppliers in advance detailing both the written policies that they wish to see and a list of daily/weekly/monthly check sheets that will prove that measurements are carried out as described in the manual (see Fig. 3.1). There is an additional opportunity to check this documentation during the factory inspection on the day when systems can be viewed in action as well as in retrospect. The existence of an ISO certified system would not necessarily ensure approval as the quality management systems required are often specific to that retailer and matters of supreme importance to them might not be covered by the certification process. For example, retailer requirements on the welfare of animals before and during slaughter may not be covered in the ISO certification process but might form a major part of a retailer's policy. Adherence to the retailer's policy would therefore be equally as important as ISO certification.

Retailers are primarily concerned with the safety of their customers. The existence of a fully documented HACCP plan is a legal requirement and therefore a detailed examination of this is essential before approval. A retail auditor may ask to see the HACCP plan for a particular product in advance of the visit as the quality of this would be a good indicator of the management's understanding of HACCP. However, the existence of a plan is not enough. During the factory inspection the auditor should specifically ask departmental managers, line leaders and operatives where the critical points in the area are and how often they are monitored. The ability to answer additional questions on why these points are critical would give a measure of the effectiveness of hygiene training. The design of quality control sheets can assist in reinforcing this knowledge by indicating where the critical control points are in the process and what action should be taken if there is a deviation from permitted values. The approval audit must include a check on the frequency of staff training together with the adequacy of training records.

A further key requirement would be the traceability of all ingredients from the coding on incoming raw materials through processing into the finished pack. The systems for doing this would be documented in the manual but can be checked during the factory inspection and/or by asking in advance for all of the records for one product made on a particular day to be available. These would then be examined during the audit. Full traceability is essential to ensure that withdrawals from sale for both safety and quality purposes catch all affected products. Sound systems are particularly necessary if the specification allows reworked material to be included in a finished product no matter how small the quantity. The practical application of the system can also be verified by checking the labelling on bins and racks during the factory inspection.

A key indicator of management commitment to quality and safety would be regular internal audits of the systems of work. These must be well documented and show that action has been taken to correct any non-compliances with the written quality management system.

The factory approval visit must be fully comprehensive as it is much harder to obtain improvement once supply has begun. Withholding approval until the required standards are reached usually results in speedier compliance. Retailers may use scoring systems before approval but extreme caution is needed with

Dear

LEGAL COMPLIANCE AUDIT

I am writing to advise you that I intend to carry out a factory audit at your premises on . . .

I intend to examine the following aspects of your Legal Compliance and Quality Management Systems and will require production of the documentation identified below:

1. Legal Compliance Management
2. Legal Compliance Systems
2.1 ☐ HACCP Charts and Records.
2.2 ☐ Product Specification.
2.3 ☐ Pest Control Records.
2.4 ☐ Metal Detector/Checkweighing Records.
2.5 ☐ Traceability Records.
2.6 ☐ Internal Audit Records.
2.7 ☐ Complaints Records.
2.8 ☐ Other.

3. Premises and Equipment
3.1 ☐ Cleaning Schedules.
3.2 ☐ Maintenance Records.
3.3 ☐ Other.

4. Raw Materials
4.1 ☐ Records relating to audits of raw materials.
4.2 ☐ Records relating to audits of raw material suppliers.
4.3 ☐ Other.

5. Process Controls
5.1 ☐ Production Records.
5.2 ☐ Other.

6. Inspection/Test/Analysis
6.1 ☐ QC Records.
6.2 ☐ Other.

7. Packaging/Storage/Distribution
7.1 ☐ Lot Marking.
7.2 ☐ Other.

8. Staff and Training
8.1 ☐ Training Records.
8.2 ☐ Health Screening Procedures and Records.
8.3 ☐ Other.

Please ensure that all the relevant personnel are available to discuss the above.

Should it be necessary to cancel/postpone my visit, you will be notified by telephone at the earliest opportunity.

Fig. 3.1 Legal compliance audit checklist.

these. It may be possible to obtain a high score while failing to control a critical point in the process, thereby putting customers at risk. Any scoring system must therefore employ a weighting procedure which takes account of the risk assessment of the product type. There is no substitute for the experience of a knowledgeable and well trained auditor.

3.2.2 Auditing a supplier's quality of service

Auditing of a new supplier should not concentrate only on legal and safety issues. The retailer will have various requirements as to capacity, shelf-life and delivery deadlines. Conversations between buyers, technical managers and potential suppliers as to volume requirements in advance of the visit are essential. Approval of a factory to supply 500 stores with a chilled short-life product, if the factory only has refrigeration and packing equipment to supply 200 stores daily, will usually end in short deliveries or poor quality. Allowance should be made for additional volume requirements for promotions and supply to other customers of the factory must be considered. The auditor should therefore check aspects such as cooking capacity, line speeds, cooling capacity, etc. at the time of the visit to ensure that the predetermined requirements can be met.

Short shelf-life products such as poultry meat may also present other difficulties. Supply will depend upon a schedule which allows for slaughter, chilling, packing, transport to depot and onward distribution giving sufficient days in store for sale. When planning to import from abroad this can cause difficulties. The auditor must consider measures that could be taken to extend shelf-life. Modified atmosphere packaging is particularly useful in these circumstances and the auditor may decide that provision of this is a condition of supply.

The quality of raw materials in use must be appropriate for the product and a knowledge of suitable ingredients suppliers is invaluable to an auditor. For example, if the factory produces fish ready meals, they should be serviced by companies with vessels that fish in areas that provide the best quality. Questions about the catchment area of supply should be included in the pre-audit questionnaire in order to eliminate visits to unsuitable potential suppliers. However, during the audit, raw material supply records should be checked to ensure that the fish is coming from the quoted suppliers. Some retailers may even specify the sources of ingredients to be used in order to ensure that their own codes of practice are adhered to at all stages of manufacture and this should also be checked during the audit. The authenticity of ingredients which are claimed to be free from genetic modification or suitable for vegetarians can be checked at this stage.

The type of packaging used by suppliers can also be critical. If supply is required in plastic bottles then a visit to a plant that can only fill into glass would be a waste of both parties' time. Sources of packaging supply, the nature of materials, e.g. recyclable, ability to withstand conventional cooking, grilling or microwaving are important, as are conditions under which packaging materials are stored and sterilisation before use, if appropriate. A certain amount of pre-visit questioning will ensure the best use of limited time.

3.3 Routine auditing: existing suppliers

Legislation requires that retailers take all reasonable steps to ensure that the products they sell are safe. There should be a specification for each retailer brand product that lists required checks to demonstrate this. Clear documentary evidence must be produced by the supplier to show that appropriate control measures are in place. It is therefore necessary to carry out regular re-visits to ensure that the management systems originally approved are still functioning and that no changes have been made without the prior approval of the retailer. The frequency of these 'due diligence' audits should be based on a risk assessment. Audits of high care factories producing highly perishable products such as sandwiches or cooked prawns should be more frequent than those of factories which produce more stable products such as rice, flour or canned goods. Before carrying out a routine audit the supplier file should be reviewed and the previous audit report checked. The new audit gives the opportunity to ensure that matters raised previously have been dealt with. If process control check sheets were found to have incorrect tolerances then the pre-audit letter should ask for more recent records of this process to be made available. These should then be inspected to verify that corrective action was taken. If training records were not up to date then the auditor should ensure that this non-compliance with approved procedure has been put right. If more frequent checks on the number of pieces of meat in a ready meal had been requested as a result of customer complaint, then these records should also be requested in advance and checked during the audit. If there was evidence that the raw material supply base was too small to ensure adequate supply or too large to ensure adequate quality control or auditing to the agreed frequency, then the records of action taken to review the ingredient supply base should be requested. If these records are requested in advance then the time taken for the audit is profitably spent and an opportunity is provided to review the management systems and ensure that they still provide the retailer with assurance that products are safe, legal and to the required quality.

Auditors new to a site could profit by concentrating on one or two critical control points to ensure that HACCP records are valid for the process and ingredients or they may decide to check all of the records for one product manufactured on a specific day. There is more than one way to complete a satisfactory review audit that will demonstrate whether or not the supplier is managing the production system well. Advice may always be sought from a previous auditor but it is essential to keep an open mind and be prepared to change the process if circumstances have changed.

3.4 Non-routine auditing

In addition to routine audits it is occasionally necessary to carry out further audits of a factory as a result of information obtained from a retailer's own quality auditing systems such as customer complaints records or random quality control tastings. The results of tests by outside sources such as the Department of Health or the Consumers Association, anonymous tip-offs or adverse media reports might also lead to an audit. These audits may be carried out unannounced, particularly if there are grounds for suspicion or a whole industry review is under way.

3.4.1 Quality audits

Retailers must have systems for checking the quality of the products sold as their own brand if they are to comply with food safety legislation. A useful first line check is often made at central distribution depots. Random opening of cases and packs by trained auditors with checks against key quality parameters might indicate that products were being delivered out of specification, e.g. tomatoes of the wrong size or degree of ripeness. Store staff occasionally report to head office that quality has deteriorated. An example of this could be poor evisceration or excessive feathers on chickens supplied for rotisseries in store. Regular head office tastings on a rota basis might detect varying viscosity of cream, sauces or other liquid products.

These variances may be dealt with initially by a telephone call to the supplier followed by more retailer checks. However, if quality fails to return to the tolerances agreed within the specification then it may be necessary to carry out a quality audit. This will usually follow a different format from the routine variety. If the product has a long shelf-life then the auditor should ask for samples from each production still in stock to be brought out for the audit. There may also be retained shelf-life samples from each production. The records for the last ten productions should be obtained and may need to be asked for in advance of the visit as they may be stored off-site. Finally, a store visit before the audit to collect samples currently on sale will allow the retailer and supplier to assess products together and identify where and why deviations have taken place.

Audits in these circumstances might concentrate on incoming raw material checks if poor quality ingredients are evident, e.g. gristle in meat pies would lead to checks on incoming meat inspection records whereas variations in the gravy viscosity would require checks on yield, cooking time and the weight of the thickener. If the product was not holding up for its full shelf-life, then the check would concentrate on looking for unacceptable delays in processing and might require a detailed traceability exercise to identify the cause. In all of these circumstances auditors should ensure that they are fully familiar with the process specification and finished product quality checks before they visit so that they can concentrate on the areas most likely to have caused the problems. Finally, it is essential that during this audit the monitoring frequencies are reviewed to ensure that potential problems are spotted and dealt with in the future and that staff are retrained where necessary on the standards that are required for the product.

3.4.2 Customer complaints

As part of their own due diligence systems retailers are obliged to analyse complaints data and act as soon as a problem is perceived. Stores will record details of the type of complaint and the codings on packs brought to the customer service desk. They will report either to a customer service centre or to the buyer or technical manager at head office. If there is a sudden spate of glass, metal or other foreign body contaminants or sickness reports from one batch of the product then these will be followed up immediately by telephone calls to inform the supplier, a decision to withdraw a batch from sale and a visit to the supplier to check procedures and records. Visits might also be carried out if complaints data collected over a period of weeks indicated that the level of complaints had increased. This trend analysis uses sales data collected from check-out scanning of long-life products or production figures from short-life products to calculate the number of complaints per million products sold. A level of ten complaints per million sold is often the target. It is essential to check the sales figures, as very successful promotions might lead to an increase in actual numbers of complaints, although when viewed against sales data the rate of complaint may be downwards.

Complaints are grouped into type, e.g. glass, metal, extraneous vegetable matter, bone, gristle, low meat content; then, if there is cause for concern because a specific type of complaint has increased, an audit will be necessary. When complaints are caused specifically by poor raw material quality, poor process control, unacceptable foreign bodies or poor hygiene, then the audit should be specifically targeted. An increase in a variety of foreign body complaints might suggest an untidy factory allowing extraneous matter to enter production areas. In this case a walk around the site collecting any small unnecessary items into a plastic bag might have the desired impact on line management. If there is an increase in extraneous vegetable matter in vegetarian products or bone complaints in meat, fish or poultry products, then extensive sampling of any bought raw materials or examination of butchering and filleting standards in-house should be carried out to identify the source of complaint. If traceability records indicate that one raw material supplier is implicated, then specific action must be agreed to reduce the level of complaint. In all cases the visit, checks carried out, and action taken must be carefully documented and records retained in the retailer's files as evidence that the auditor has carried out all reasonable checks as part of the retailer's duty to preserve product integrity.

3.4.3 Adverse microbiological trends

For many types of products there will be a microbiological specification. The retailer and supplier will have agreed the number of samples of each product to be analysed on the day of or day following production and throughout the shelf-life. The standards applied are usually based on legislation or good manufacturing practice (GMP). The organisms tested for will vary depending upon the nature of the product but will usually include a total viable count of organisms per gram. This will give a measure of the effectiveness of any heat treatment or processing procedures, an indication of expected shelf-life and possibly a measure of the quality of ingredients used. There will be a check for pathogenic micro-organisms of the type or types most likely to be found in that product group depending upon the raw materials used and the degree of handling by staff. The specification usually has three levels – a target based on GMP, a higher level at which the retailer should be informed on the day the result is obtained, and a withdrawal level at which the product will be deemed to be unsafe, unacceptable on the day of production or unlikely to be sound at the agreed end of shelf-life. If any product figures in this last category the supplier must immediately telephone the retailer and arrangements will proceed to withdraw stocks from sale. However, if results of microbiological checks fall within the first two categories they will be reported to the retailer using some form of graphical trend analysis. If these results show rising total viable colony counts on the first day after manufacture, then it is possible that eventually some would not be acceptable at the end of life giving rise to customer complaint. The retailer and supplier therefore have a duty to monitor these trends and take action if deterioration in quality is predicted. An audit in these circumstances should concentrate on the microbiological quality of incoming raw materials to ensure that they are sound, and should check on any variation in processing or storage times to ensure that micro-organism levels are being reduced to a minimum and that there is effective control of the chill chain if this is relevant. Again, action must always be agreed and documented at the end of the audit and immediate re-checks agreed. In addition the frequency of checks should be reviewed to ensure that the problem can be prevented from recurring.

If cleaning procedures within a factory have been inadequate, then this can usually be detected by monitoring coliform or Enterobacteriaceae counts in products on the first day of life and a level at which the retailer is informed will often be agreed within the specification. When trend analysis shows levels are rising, or several results above the reportable specification limits are received in one month, then a hygiene audit should be carried out. This should concentrate on the cleaning methods, frequencies and materials in use. A check on the dosage level and residence time is essential as new or inadequately trained staff could occasionally be the cause of the problem. These audits should therefore be carried out during the cleaning shift and the supplier's hygiene manager should be present. Swabs or ATP kits can be used to measure residual contamination and dead spots must be identified so that cleaning methods can be improved to achieve the desired results.

3.4.4 Adverse media reports

In the past ten years there have been several reports in the Consumers Association magazine, *Which*, and surveys carried out by local authority trading standards officers on behalf of the Ministry of Agriculture, Fisheries and Food, the Food Standards Agency and the Department of Health on all types of products. The results of these studies are frequently reported in the mainstream press. The studies usually select specific groups of products which are then analysed and the results compared to current legislation or specific quality standards. Examples of this type have been meat content of sausages and water content of hams. When retailers are identified as selling products below acceptable quality standards then of course they will wish to protect their reputation by carrying out a thorough investigation. The analytical results may be affected by many factors, so in addition to checking processing methods and records the auditor should if necessary be accompanied by a company analyst or consultant with specific expertise in this area. There have also been reports on pesticide residues in fruit juices where the input from a horticulturalist during the audit would be invaluable. The report on Listeria sps in cooked chicken and pâtés caused a flurry of activity and many retail technical managers and suppliers were ill-informed. This was therefore an occasion when an audit by a food technologist accompanied by a microbiologist would have been most helpful. On these types of occasions a delayed audit would enable the best informed people to be assembled by both the retailer and supplier so that the audit can examine all contributory factors thoroughly and ensure that more complex control methods are put in place for the future if this is found necessary. However, unannounced visits should also be considered if spot checks on current activities are deemed necessary.

Most retailers occasionally receive anonymous telephone calls from people who describe totally unacceptable conditions or practices within premises producing own brand products. The calls are usually related to poor quality ingredients, dirty equipment, abuse of agreed quality parameters or even the use of child labour. An auditor who knows the supply base well will have spotted indications of this during their routine work monitoring quality, shelf-life and customer complaints. However, in these cases an unannounced visit may be the only way to elicit the truth. Before carrying out a visit in these circumstances the auditor should carefully consider where and how they are most likely to find the evidence to substantiate the allegations and prepare a check sheet so that they are thorough in their investigation and cannot be diverted from their task by a determined supplier. If the auditor does not know the supplier, premises or processes well, then this pre-preparation is essential and, if deemed necessary, two people should carry out the visit. The audit must be carried out at the most appropriate time of day, night or weekend, i.e. when non-conformances to agreed procedure are most likely to occur. However, the auditor should bear in mind that occasionally complaints are made by disgruntled employees and could be malicious; therefore it is important to keep an open mind.

3.4.5 After product withdrawals

Rarely, mistakes occur and products must be withdrawn from sale for either safety reasons or because of unacceptable quality defects. There is then a need to audit suppliers to ensure that control systems are in place to prevent that type of incident from recurring. The timing of this audit will depend upon the seriousness of the incident, the past history of the supplier and the auditor's knowledge of and confidence in the management. If there is a serious safety or quality issue then all further production will be halted until an audit is carried out. In this case an immediate audit is necessary unless an alternative source of supply is available. If the retailer has another supplier who can pick up the required volume immediately, then the audit can be delayed for 48 hours or up to a week.

Rejection of a product from a small factory with limited resources would require a rapid visit to ensure that the retailer has confidence that corrective action has been taken. Indeed, if the problem has been caused by an ingredient, then the retailer might have more knowledge of alternative sources of supply than a small manufacturer and their intervention would be appreciated. However, if the same type of incident occurred in a product from a large company with extensive technical resources, they might be able to solve the problem themselves and the audit could be delayed until alternative ingredients have been sourced. The audit would focus upon why the defect was not spotted before withdrawal became necessary and discussion would centre on raw material inspection, handling policy and records.

When incidents have serious safety implications such as the undercooking of a meat product, then an immediate audit is called for as there must have been failure to control a critical point in the process. The retailer will need to be reassured that this cannot happen again before production can resume. Overcooking of meat, however, could be discussed on the telephone and the visit planned for a later date. The retailer would then require process control records for several production cycles and retained samples to be made available and the audit would focus on the effect of varying cooking times on eating quality. Finally, if the auditor has confidence in the management and the problem is complex, then a delay in carrying out an audit would allow the supplier to spend several days investigating the incident and the audit would then concentrate on reviewing systems and frequency of checks. Agreement would then be reached and documented showing a course of action to be followed in future.

3.5 Summary

Audits carried out by retailers or their agents could be regarded as a chore undertaken to fulfil legal requirements. Completion of a pre-designed tick sheet might be seen as the whole purpose of the audit by a technologist with little understanding of a retailer's requirements. However, if both the retailer and supplier prepare in advance and approach the audit in a spirit of co-operation, then the audit provides an opportunity to review procedures, change monitoring frequencies and eliminate tasks no longer required. In other words, the audit becomes an opportunity to move both businesses forward to mutual and customer advantage.

4

Regulatory verification of safety and quality control systems in the food industry

V. McEachern, A. Bungay, S. Bray Ippolito and S. Lee-Spiegelberg, Canadian Food Inspection Agency, Nepean

4.1 Introduction: the role of safety and quality control systems in the food industry

Over the last fifty years there has been a great deal of change in the food industry. The number and variety of food products available to today's consumers have increased dramatically, along with their attitudes to and expectations of food and food products. Only a few decades ago, the majority of consumers purchased their fruits, vegetables and meats at the local grocery store and prepared their meals at home. In most cases these food products were produced in local processing plants that received their raw materials from regional farm and fish industries. These products had limited distribution and were produced in volumes sufficient to satisfy local and regional needs.

Today, consumers lean toward convenience, purchasing microwave-friendly foods, ordering takeout and eating at restaurants. But while technological advances in food processing and the modernization of transportation and food distribution mechanisms have increased the variety of the foods we eat, they have also presented new hazards and concerns that must be addressed. Outbreaks of food-borne illnesses now have the potential of being national, continental or even global in scale.

As the food industry has evolved, so has the government's approach to food inspection. The first Canadian Fish Inspection Act was enacted in 1919 to address the fraudulent activities of unscrupulous fish traders that threatened the reputation of Canadian fish products in European and New England markets. Over time, food inspection programs became comprehensive and multi-faceted, addressing not only fraudulent practices, but also food safety and quality. Government resources were concentrated in the food processing plants which provided an ideal opportunity to inspect the processing conditions and the final products prior to distribution.

These inspections were based on a traditional approach that focused on "snapshots" of the processing plant environment and comprehensive sampling and inspection of the final product. The plant processing environment was evaluated against prescriptive construction, sanitation and hygiene requirements. The product inspections included label evaluations, organoleptic analysis, microbiological analysis and chemical analysis. Laboratory analyses were carried out to detect a limited number of chemical, physical and biological indicators of unacceptable products or processes. Under this traditional inspection approach, a paternal relationship between the inspector and the processing plant would develop where the inspector identified the problems and the plant then took corrective action. In many cases, the inspector fulfilled the role of quality control and because the system relied heavily on final product inspection, mistakes were not identified until they were already processed into the product.

In the 1980s, the international marketplace began making greater demands on the food processing industry and government food inspection agencies to provide assurances that food products were nutritious and safe. These demands were prompted by informed and knowledgeable consumers who expected zero risk from their food products. As a result, importing countries started to request government certification of an increasing number of food products. Under the traditional inspection program this meant more inspections. Unfortunately, the additional resources required to perform more inspections were not made available.

New or re-emerging pathogens have also meant a re-evaluation of some of our previous views and accepted approaches to controlling food-borne illnesses. The emergence of new pathogenic agents such as E. coli 1057:H7 has changed the production and process controls for a wide variety of products. Recognized pathogenic micro-organisms are now being identified in new vehicles and products that they formerly were not linked to.

All of these factors have forced industry and government to look for new and innovative ways to achieve safe and wholesome products. At the same time, government regulatory agencies are also facing new challenges associated with a complex, globalized and technology-driven food industry. Industry has realized the benefits from developing effective food safety and quality control systems, and regulatory agencies have had to develop more efficient techniques to verify the effectiveness of industry's programs.

4.2 The principles of an effective food safety and quality control system

4.2.1 Food safety versus food quality

The goal of all food processors is to produce a product which meets the expectations of their customers, thereby creating or maintaining a market niche.

When articulating their expectations of food, most consumers will indicate a preference for a "quality" or "high quality" product. Quality, however, is a very subjective term, meaning different things to different people. In fact, the word "quality" can refer to any product attribute and is often avoided when discussing food safety due to the confusion with sensory or grading standards.

The ISO definition of quality is the totality of characteristics of a product that bear on its ability to satisfy stated and implied needs (ISO 8402:1994). Or, in other words, to be considered a quality product it has to meet the expectations of consumers. Safe food can be defined as a product which, when prepared and consumed under its intended use, will not present a health hazard. Consumers expect safe food products, as well as accurate labelling on all the product they purchase. It could therefore be stated that food safety is a subset of quality. This is not to undermine the importance of food safety as it is and always will be the foremost quality requirement of the consumer.

The production and sale of safe products is usually regulated by governments in the applicable jurisdictions. In addition to safety, many other requirements must be met before a product can satisfy all regulations and be permitted market access. For example, fish – which is processed for export in Canada – must meet specific sensory standards. There can also be requirements for packaging and labelling. Many countries have particular requirements for imported food products and most food companies have very exacting specifications for their suppliers. All of these factors must be considered when a processor decides to produce a "quality" product to meet their customers' expectations. It is in this context that the term quality is used in this chapter.

4.2.2 Industry control systems

Before commencing to produce any product, processors must be aware of consumer and government expectations, including those related to food safety, sensory characteristics, other regulatory requirements and specific customer or foreign country requirements. Once these expectations are known, the processor can design a control system which provides a reasonable level of assurance that they will be met.

This system is normally defined by industry in terms of quality control. This refers to the operational techniques and activities that are used to fulfill customer expectations. The quality control procedures used will vary greatly from processor to processor and from sector to sector with regards to complexity, level of detail and scientific basis. To ensure that an appropriate control system is developed to reflect the product and processing conditions, it is recommended that a "systems approach" be taken. A systems approach requires that the system be developed with: the full support of management; participation from all levels of production; decisions which are recorded and based on scientific principles; and well documented procedures and activities which are made available to all staff responsible for carrying out the activities.

4.2.3 Effective food control system

An effective food control system, developed following a systems approach, must include certain elements which will lead to the development, implementation and maintenance of the system. These elements are common to all quality systems and can be summarized in two different categories – system development and system support.

System development

The development of a quality system involves three basic components: know what standard has to be met; identify control measures that will be used to meet the standard; and know what corrective action procedures are to be followed when the standard is not met.

Identification of the standard

The first step in developing a control system is to identify what standard must be met to produce a quality product. A standard in this context can be defined as any measurable attribute that would differentiate between an acceptable product or process and an unacceptable one. To determine these standards, the processor must first understand the products and processes that are covered by the quality system. The processor should map out the processes and list the products produced. For each product, the processor must identify those product attributes which need to be controlled. For each attribute, the standard to be attained must also be identified. Standards can be a regulatory requirement, a product specification, a processing condition or even a customer demand. For example, customers may prefer to purchase cod fillets which are greater than 250 grams.

Identification of the control measures

Once these standards have been identified, the processor must then determine how they will be met during normal production. The processor will establish control measures that will be used to monitor the product or process at an appropriate frequency for conformity to the standard. In addition to the frequency, the processor must identify the person responsible for monitoring each control, the actions to be taken (or how the monitoring will be conducted) and any records that must be kept. Control measures include having specialized processing equipment, training employees to perform certain functions and devising procedures which are well documented and accessible to all responsible employees. It is essential that monitoring be conducted for measurable attributes on a real time basis so that action can be taken immediately when deviations are detected. Using the previous example of a standard for cod fillets greater than 250 grams, there are many options available to the processor. The use of technology and/or specialized equipment to fillet and trim the cod to this specification may be chosen. There may also be continuous monitoring through an automated check weigher which records the weight of the fillets on line. Or the processor may prefer to have trained filleters and trimmers, and a regular quality control (QC) check on the weight of the final product. In either case, the processor must determine which control measures are appropriate for the operation with consideration given to the resources available, the volume of production and any other challenges which may be faced in maintaining the required standard.

Identification of the corrective action procedures

When the control measures indicate that the product or process is not meeting the standard, the employee conducting the monitoring must be aware of which actions should be taken to correct the deviation. As part of the quality system, the processor must identify corrective action procedures to be followed when this situation arises – procedures deemed appropriate to the standard that is being controlled. As well, the level of detail stated in the procedures should be related to the seriousness of the deviation. Continuing with the example of the 250 gram cod fillet production, the processor may identify a corrective action procedure requiring that any production since the last monitoring be held separately for further compliance sampling and, if necessary, the repackaging of the product as per specifications. For health and safety concerns, the corrective action procedures would be more direct and may result in the holding and destruction of any implicated product. The processor should also consider the use of operational standards to act as a buffer to the actual standard and reduce the need for product action. In this case, an operational standard could be 265 grams, so that when monitoring indicates a weight below this limit then adjustments can be made in the equipment or direction given to employees.

When developing the corrective action procedures, long-term solutions should also be considered. These actions should be based on the information gained from the deviation and are intended to prevent the recurrence of that deviation in the future. For example, a processor may want to retrain employees or even replace a certain piece of equipment that is used for the control measures.

System support

Once a system is developed, its implementation and maintenance require a series of support elements to ensure that the system is effective and is being implemented as written. These elements can be summarized as follows.

System validation

Before a quality system is implemented, the processor must validate that the standards have been identified and control measures have been set in place. The method of validation must be appropriate to the standard and controls, and may involve several approaches, including: the use of regulations, scientific literature or accepted industry practices; or conducting studies either in-house or through an accepted authority. Depending on the situation, a processor may use one or a combination of these methods. If a control measure is new and deviates from what is considered common practice or if it involves a health and safety concern, then further work may be required. For example, the processor may want to have a controlled production line with data collected at a specified period of time to validate the controls or standard. There are many consultants or other authorities who can conduct these studies should the processor lack experience and/or technical capability. The processor must also consider regulatory requirements for this component, the fact that permission to conduct test production studies on some products may be required, and – for certain procedures such as thermal processes – that a process authority may be mandatory for validation.

Control verification

Processors should build an extra level of controls into their quality systems to verify that the control measures are being implemented as written. It is important to verify the system continuously to catch potential problems before they affect the final product and to give management confidence in the system. The verification procedures can be tests or measurements which are more detailed and are not required to produce immediate results as is the case with the control measures. A verification procedure for the production of 250 gram cod fillets could include a periodic sampling of final product by QC to verify the weights. Other such procedures include verification of records on a regular basis, calibration of equipment on a preset frequency and a complete systems review by management. This final verification procedure is extremely important as after the initial systems development and implementation changes to processing conditions or desired product attributes will occur over time. A complete review is recommended at least once per year to verify that the current standards are being applied, the procedures are effective and to recommend potential improvements to the system. Processors should constantly be striving for improvement in their system and an annual review is an excellent mechanism to facilitate this.

Record keeping

Effective record keeping is essential in any quality system to provide evidence that the system is being implemented as written for regulatory agencies, customers and for internal verification. It is also important to demonstrate trends before a problem arises or to lead to potential improvements in the system. There are two basic types of records which apply to a quality system, the record of the development of the quality system and the records taken as a result of the implementation of the quality system.

The development of quality systems normally includes support and input from many individuals and usually takes place over a considerable period of time. During this phase, there are numerous decisions taken and authorities referenced. This information should be captured, not as a part of the quality system, but as a separate record. This information is essential to justify, if necessary, to regulatory agencies or customers why certain actions or activities have been included and also to assist in the future development and evolution of the plan.

During the development of the quality system, the processor has to identify what is recorded as a result of the activities described in the plan, including monitoring, verification, corrective action and employee training. It is important, however, to balance the volume of record keeping with the true needs of the organization and the resources available to deliver the system. Some record keeping will be regulated as a mandatory requirement; others will be required for customers or the use of management. These factors must be considered and priorities set on what records will be kept because any system developed with record keeping requirements beyond the capabilities of the organization will be destined to fail. Consideration can be given, in some areas, to record keeping by exception – that is, where a record is not taken unless a deviation from the standard is identified. In such a case, a note of the deviation and the corrective action taken is all that is recorded.

Other factors which can influence the effectiveness of record keeping include ensuring that employees understand why records are being taken and how this task can best be performed. Inaccurate or illegible records can do more harm than not taking any records in some cases. Processors should also try to simplify their records, eliminating the gathering of information which is not used or required and, where possible, seeking to combine records. Consideration can also be given to the use of technology to allow for continuous monitoring or automatic capture of data through computers or remote sensors. The retention time for records is a very important issue and again must be balanced with regulatory requirements and available resources. Records should be retained for a period of time which is relevant to the shelf-life of the product. They should be stored in a manner which is easily accessible (as may be required in the case of a product recall) and also in a manner which will maintain product integrity. The computer can also be an effective tool for this task; however, proper attention must be given to security and backup of data. Again, there may be specific regulatory requirements for record retention and this should be verified when developing this aspect of the plan.

System maintenance

It must be understood that a quality plan should not be regarded as a static document but as a document which is continuously being improved and developed. Over time, it must change as better techniques for monitoring or more efficient controls are identified, product requirements change, or additional hazards emerge. These provide excellent reasons why an annual verification is necessary, and when these cases are identified, why a system must be modified. During modification, it is important to capture what was changed and why. This now becomes a record of the system development and will aid regulatory agencies and customers when conducting a review of the system. Once a modification of the system has been made, it is essential to have a process in place to update the earlier versions of the plan and to inform any employees whose activities may be affected by the changes.

4.2.4 HACCP as a food control system

The most well known systems approach to food safety is the hazard analysis critical control point (HACCP). The HACCP concept was originally developed by the Pillsbury company to meet the demands of NASA for safe food products to be used during manned space flights. To produce these food products, a system was required which went beyond the traditional methods of sampling and analysing finished products for the presence and levels of identified hazards. In developing their system, Pillsbury took the approach that if it was understood what makes a food unsafe then control measures could be developed to prevent the hazards from occurring and reaching the consumer.

Over the last decade the application of HACCP has become an internationally accepted standard for the control of food safety hazards. Many countries now require that foods which are processed, imported or offered for sale must be processed in a HACCP environment. The application of HACCP is normally described in terms of seven principles which have been formalized by groups such as the Codex Alimentarius of the Food and Agriculture Organization of the United Nations. These principles are as follows:

1. Conduct a hazard analysis.
2. Determine critical control points (CCPs).
3. Establish critical limits.
4. Establish monitoring procedures.
5. Establish a corrective action system.
6. Establish verification procedures.
7. Establish documentation and a record keeping system.

The principles of HACCP are similar to the elements of the quality system described in the previous section. In fact, HACCP is really only a system approach designed specifically for food safety with a formalized hazard analysis at the front. A detailed explanation of the application of the HACCP principles need not be given in this section, as this can be found in many references and documents (including Codex Alimentarius alinorm 97/13 A). However, a brief overview is given for each principle for comparison with the elements of an effective quality system. This comparison is useful to visualize how HACCP can fit into a processor's existing quality system.

Principle 1: Conduct a hazard analysis

The hazard analysis follows a scientific approach to ensure that the processor has identified all significant biological, chemical and physical hazards that may be introduced or that can grow to levels which may present a human health risk in the final product. There are many different techniques for conducting a hazard analysis. Most, however, follow the basic sequence of mapping out the processes and brainstorming, using a team approach to identify all potential hazards at each process step. Each hazard is then considered in relation to the severity of the potential illness and the likelihood of occurrence or risk. When hazards are identified as significant then appropriate controls must be put in place.

Principle 2: Determine critical control points
Significant hazards are not necessarily controlled at the process step in which they were identified. This principle is intended to determine at which steps a control should be applied to prevent or eliminate a food safety hazard or reduce it to an acceptable level. If a process step is determined to be the most appropriate for control purposes, then the step is considered a critical control point (CCP). Examples of CCPs are cooking steps, addition of additives or metal detectors. A tool which is very useful for determining CCPs is the Codex decision tree.

Principle 3: Establish critical limits
At each CCP, the processor must determine the value which separates a safe product from an unsafe one. This value can be a temperature or time which must be achieved to ensure destruction of a pathogenic bacteria, a certain pH to prevent the growth of bacteria, the level of a preservative or the size of detectable metal particles.

These first three principles relate to the identification of the standard element of quality system development. The HACCP system, however, should be based on a more scientific approach and relates only to food safety.

Principle 4: Establish monitoring procedures
At each CCP, the processor must establish monitoring procedures to determine that the system is operating within the critical limits identified in the previous principle. The monitoring procedures must indicate what will be monitored, how the critical limits will be monitored, how frequently and by whom. It is important to have monitoring procedures which produce immediate measurable results on which action can be initiated, since there may be potential food safety implications. This principle is the same as identification of control measures in the quality system development.

Principle 5: Establish a corrective action system
When the monitoring indicates that the critical limit was not met, then corrective action procedures must be initiated. These procedures should be identified in advance so the employees who are conducting the monitoring will have direction on the steps to take when a deviation is identified. This is essentially the same as the three elements of the quality system development outlined above.

Principle 6: Verification procedures
When controlling food safety hazards, it is extremely important to have an additional level of control to ensure the system is operating as it was designed. Processors have to identify activities in addition to the monitoring – but undertaken on a less frequent basis – to review the implementation of the plan through the records or through additional tests or analysis. This principle could be considered as part of the control measures or as one of the system support elements.

Principle 7: Establish documentation and record keeping system
The last principle of HACCP is to establish a documentation and record keeping system. The guidelines presented in the systems support component of quality systems above pertain to HACCP record keeping. However, because the HACCP plan controls food safety concerns, the importance of effective record keeping should be stressed even more. In the case of an illness, accurate records may be a processor's best defence to prove to the authorities and the public that his or her products are safe or that there is control. The presence of specific regulatory requirements for record keeping of food safety controls should also be considered.

In comparing quality systems and HACCP, it is obvious that the two systems are very similar. If a processor wishes, the development and implementation of a HACCP system can be made part of a processor's overall quality system thus leading to the integration of control activities and the creation of efficiencies in the delivery of the quality system. This should allow a more effective utilization of resources and a greater concentration on the most important control activities, including food safety.

4.3 The role of government and industry in achieving food safety and quality

The goal of achieving a safe and wholesome food supply is a shared responsibility among all stakeholders along the gate-to-plate food continuum and includes government, industry and consumers. The system is only as strong as its weakest link and, therefore, it is important that each partner understands and carries out their responsibilities.

Each stakeholder along the food continuum is responsible for their impact on the product and must practice due diligence to ensure that food safety and wholesomeness is maintained while the product is under their control. The combined efforts of each and every industry sector, from the feed manufacturer to the processor on through to the distribution and retail sector, are fundamental to achieving this goal. In order to practice due diligence, stakeholders must be knowledgeable about the potential food safety hazards and the regulatory requirements which pertain to the product and processes for which they are responsible. With this knowledge, stakeholders can design and implement effective control measures to prevent and eliminate potential food safety hazards and ensure that product meets regulatory requirements. By using the systems approach described in the previous section, industry can demonstrate that due diligence has been practiced and product integrity maintained while under their control.

The first role of government in achieving safe and wholesome food is to set food safety and regulatory standards through food legislation. These standards provide the foundation of the food inspection program, and in order to facilitate trade in food products, they must be consistent with international guidelines (such as those developed by the Codex Alimentarius Commission). The standards must also provide the framework for industry to develop and implement quality management systems. This can be accomplished by drafting reference standards, based on regulation, that describe the goals and outputs that a quality management system must fulfill.

Government also plays the primary role in assessing the effectiveness of industry's controls and level of compliance in relation to its products. As mentioned in section 4.1 above, governments have historically relied on an inspection approach involving snapshot plant and product inspections to assess industry compliance. As we move toward a systems approach to food inspection, the government's role changes from inspecting specific production lots and processing conditions for compliance on a specific day, to assessing the effectiveness of industry control measures in achieving food safety and regulatory compliance. Under this approach, government inspectors have many tools to assess the effectiveness of industry controls. For example, records of production, control measures and corrective actions gathered by industry over a period of time can now be reviewed by an inspector. In addition, traditional inspection techniques can be used to focus on areas where a non-compliance is identified or simply to verify that the control measures are effective. In effect, the systems approach does not discard relevant traditional inspection methods but continues to use them where appropriate. And it allows the government's assessment of industry practices to be more comprehensive, flexible and responsive to change. Finally, the systems approach enhances the ability of government food inspection agencies to direct resources based on the level of risk of a product and historical levels of compliance.

In addition to inspection, government food inspection agencies are also responsible for taking the appropriate enforcement action when a non-compliance is identified. This can range from detaining products for further analysis to the removal of a processor's right to operate. Whatever the enforcement action, it must be fair, predictable and equitable.

At the end of the gate-to-plate food continuum lies the consumer – the final stakeholder in the effort to achieve food safety. Food produced and delivered to consumers under an effective integrated food safety control system can still pose safety problems if the end user handles the product irresponsibly. Consumers have the right to be informed about the food they eat and their own responsibility to handle food properly. Government and industry must work together to keep consumers informed and educate them on basic hygienic food handling practices.

4.4 Regulatory verification versus audit

As government food inspection agencies and the food industry implement food inspection programs based on systems such as HACCP, the government's role will shift from inspector to auditor. This transition will pose some significant challenges to an organization that has evolved under the traditional approach to inspection. In order to get a clear understanding of what these challenges are, an examination of audit principles against the traditional methods of inspection would be of benefit.

An audit is defined as "a systematic and independent examination to determine whether quality activities and related results comply with planned arrangements and whether these arrangements are implemented effectively and are suitable to achieve objectives" (ISO 8408). To paraphrase, it is an assessment of a quality management system to determine if it is doing what it says it is doing.

The audit follows a systematic approach. Before it commences the scope is agreed to by all parties, the specific elements of the quality system that will be audited are identified, checklists are prepared and reviewed by the audit team, specific audit activities are designated to each member of the audit team, and the date and duration of the audit are set and agreed to by both the auditor and auditee. In order for an audit to be credible it must be performed by an independent and impartial party, and the audit results and conclusions must be based on objective evidence that is verifiable. The audit must measure the quality system against a defined and agreed upon reference standard and be performed by a competent auditor and audit team who are trained and experienced in the sector being audited.

There are two principal activities carried out under the audit. The first is a quality system audit of the company's documented quality system. This is sometimes referred to as a desk audit as it involves mostly a review of the documented system against the agreed upon reference standard. The second component of the audit is referred to as the compliance audit. The compliance audit is carried out once it has been established, through the quality system audit, that the company's system meets the reference standard. The compliance audit focuses on the application of the quality system and verifies that the company follows the control procedures as described in their system.

At the conclusion of either the quality systems audit or the compliance audit, the auditor and the audit team prepare a report which identifies the non-conformities that were observed during the audit. The auditee is then required to prepare a corrective action report describing the actions (what, when and who) that will be taken to rectify the non-conformities. The corrective action report is reviewed by the auditor, and either accepted or returned to the auditee for amendments. Once the corrective action has been completed, the auditor will determine if a follow-up verification is necessary to check if the corrective action has been completed and if it is effective. Once all of the non-conformities have been satisfactorily dealt with by the company, the auditor closes the audit.

The audit approach is a very effective method of testing and challenging a company's quality management system. In many cases, the audit is performed at the request of the company and provides an opportunity to have an independent and knowledgeable third party assess their quality system in a non-adversarial environment. The results of the audit are seen as opportunities to strengthen the quality system – a necessary step in the cycle of continuous improvement. The relationship between the company and the auditor does not extend outside the scope of the audit. The auditor's only role is to measure the application of the quality system against the identified reference standard and assess the degree to which the quality system is respected in the day-to-day operations of the company.

When comparing the relationship between a government food inspector and a food processing company with the relationship between an auditor and that same company, there are some significant differences. The client of the regulatory inspector is the consumer, whereas the auditor's client is the company. The majority of assessments performed by the government inspector are not at the invitation of the company, and at the best of times, may be considered by the company to be a distraction and an annoyance. Although government inspectors are very knowledgeable of the industry they inspect and base all observations on objective evidence, the non-adversarial environment is stressed through their role as regulators. Inspectors are obliged, by the nature of their mandate, to act when a non-compliance with regulations is identified. This is a very important distinction between the government inspector and the auditor.

The auditor does not have any responsibility or authority to deal with health and safety hazards that may be generated by an ineffective quality system. The auditor identifies the non-conformity and then passes the responsibility to deal with the non-conformity over to the company. Conversely, if a government inspector identifies a non-conformity that violates regulatory requirements or has the potential to generate a health and safety risk to the consumer, the inspector is obligated to take immediate steps to protect the consumer. The government inspector will take the appropriate action depending on the seriousness of the non-conformity and may require that product be detained or recalled, or that the company's food operation be suspended. The government inspector plays an important part in the food control system designed to protect the consumer. In comparison, the auditor's role, although it has a positive impact, does not carry an equivalent level of responsibility, authority or liability.

The systems approach to assessing industry compliance requires that government inspectors adopt auditing methods and techniques – as the Canadian food inspection system maintains inspectors as the primary assessor of industry compliance. The Canadian Food Inspection Agency (CFIA) has also developed a new approach to assessing the fish processing industry operating under quality systems which is referred to as "regulatory verification."

Regulatory verification applies a combination of audit and inspection techniques in assessing industry compliance and reacting to regulatory non-compliance. This approach provides government inspectors with both audit and inspection tools to assess the effectiveness of industry's controls. Auditing techniques such as analyzing and verifying industry's documented controls, reviewing records and corrective actions, interviewing company employees carrying out activities, and observing the application of in-plant control activities are used by inspectors when performing a regulatory verification. In addition, traditional inspection techniques can be used to focus on areas where a non-compliance is identified or simply to verify that the control measures are effective. The regulatory verification approach is a comprehensive in-depth assessment of industry's controls and outputs that applies to both audit and inspection methods. It allows the inspector to evaluate information and data gathered over time by the company, to perform inspections of the product and the plant environment and to focus his or her verification efforts based on risk and compliance. This allows the decisions of the inspector to be based on a greater amount of information than in the past and gives more flexibility in assessing industry controls.

4.5 Regulatory verification of industry food safety and quality control systems

The term "regulatory verification" describes a set of activities which are carried out by, or on behalf of, a government regulatory body such as the CFIA, to assess compliance of a company's food safety and quality control system (FSCS) to a specified reference standard.

The reference standard defines the requirements that industry's FSCS must meet and should be based on objectives for food safety and the applicable regulations.

Regulatory verification includes both inspection and audit activities conducted in order to challenge and confirm (or deny) that industry controls are well developed, correctly implemented and effectively maintained.

For businesses that have a documented FSCS, the regulatory verification consists of three distinct phases, the industry self-verification, the assessment of the written system and the assessment of the implemented system. Regulatory verification is conducted using a team approach. This does not necessarily mean that a large group of individuals are required to assess an industry control system. However, it does reflect the need for a range of talents and a diversity of skills and knowledge. This team can be available in person or via a network of available expertise. The team leader will determine the needs for the regulatory verification activity and involve other personnel, as required, to ensure the team has the necessary knowledge and skills to conduct the verification.

Throughout the regulatory verification process, good communication with industry is emphasized.

4.5.1 Industry self-verification

Prior to the implementation of a food safety control system, businesses should conduct an in-house assessment to determine whether all required components of the system have been addressed. This is called industry self-verification.

In order to assist industry in this step, regulators can develop checklists based on the regulatory requirements. Self-verification checklists should be concise and easy to use. Checklists can be used by the companies to plan the development of the FSCS, as well as a final checking process for completion.

The process for industry use is fairly simple. Upon completion of the FSCS, a business member – preferably independent of the system development team – should use the checklist to verify that the components are all present prior to submission of the documented plan to the regulatory agency for acceptance. The checklist would then accompany or precede the FSCS submission to the regulatory agency.

Some may regard the industry self-verification as a "paper exercise" but this is a false perception. From the viewpoint of the food industry, the self-verification checklist is a useful tool for development and completion. For regulators, the industry self-verification process promotes industry ownership of the FSCS. The self-verification checklist can be an important communication tool, and regulators should design the checklist to provide the relevant industry progress data that is needed for government planning in this type of initiative. For example, checklists should be created in order to determine if the industry FSCS is sufficiently developed to proceed to the next stage – the system verification.

4.5.2 System verification

The objective of the system verification is to assess the written FSCS against the reference standard to determine if the written document is complete, and technically and scientifically sound.

The system verification is done on the first written FSCS and on subsequent amendments. The process of system verification can be discussed in several distinct stages: planning the system verification; conducting the system verification; and communicating the system verification results.

Planning the system verification

In order to assess and acknowledge any written FSCS, it is essential that a set of criteria be prepared which will serve as the basis for assessment. The criteria must be developed from the reference standard and should interpret the requirements of the reference standard for the written FSCS. The criteria can be developed in the form of a checklist and assessment report to be used by the regulator. Every effort should be made to balance ease of use of the report with the need for sufficient criteria to conduct a comprehensive assessment of the written FSCS.

The system verification is a well organized process. Regulators will not begin a system verification until the industry self-verification indicates that the written FSCS is complete and ready for assessment. It is important to remember that each FSCS is unique, and the regulator must remain open to different approaches in meeting the system verification criteria.

Conducting the system verification

In conducting the system verification, the process begins with the establishment of a system verification team, including technical or scientific resource people when required. It will be necessary to have all relevant documents available (i.e. the FSCS, the system verification report, applicable product hazards references and any relevant industry guidance documents).

The system verification is a paper review of all major components of the FSCS. It includes the prerequisite programs and flow diagrams, hazard analysis, control measures, corrective action plans, etc., but would not include very detailed procedures such as specific work instructions. This checklist is used to assess the completeness and soundness of the written FSCS. In making this determination, it is essential that the regulator consider the interaction between various parts of the FSCS, for example the prerequisite programs and the HACCP plan. These sections cannot be assessed independently of one another.

A common difficulty in conducting the system verification is determining the level of detail that should be satisfied to find the written document complete and sound. It is important to differentiate between "complete and sound" and the ideal written FSCS that individual regulatory personnel may envisage.

The system verification should be closed when the regulator has sufficient evidence to believe the written program is complete and sound. The system verification does not attest to the effective implementation of the FSCS, only that the system is found to meet the requirement of the reference standard.

It should be expected that it will take time for industry to create an effective documented FSCS. Specifically, time to analyze and develop the first draft of the written system, to test the procedures through implementation and to revise the written procedures, at least once but more likely over several system generations. The regulatory system should facilitate, and even encourage, this test and revise approach.

Communicating the results

The company should be provided with a written report indicating any deficiencies found in the written FSCS. Regulators are cautioned to avoid the opportunity to suggest ways to revise the written system for two reasons: industry authorship of the document will improve its workability; and if regulators are involved in developing the system, it may create conflict when the system is being assessed during implementation.

4.5.3 Compliance verification

The compliance verification is the on-site assessment of the implemented FSCS. It has a twofold objective: to verify that the FSCS is implemented as written and to ensure that the system is effective in meeting the requirements set out in the reference standard. It also has planning, opening, investigating and closing phases.

The compliance verification is based on audit principles. The process is designed to conduct a meaningful assessment in an efficient manner. In order to accomplish this, regulators are tasked to evaluate a "thin slice" of the company system (i.e. a comprehensive, but narrow, review of the company's control system). For example: for a large food production facility, a regulator could limit the compliance verification to an in-depth investigation of the controls in place for only one of many products produced.

Planning the compliance verification
The first step in the compliance verification is the development of a compliance verification checklist. This is prepared using the company's FSCS and in advance of the on-site verification.

Checklists are devised to address the scope and objectives of the verification. The checklist creates a guided structure for each assessment process. Depending on the nature of the verification, the checklist may take on different, yet similar shapes. Lines of inquiry for the checklist are set up using the reference standard and its structure. The checklist should contain specific activities to be conducted to test the application and effectiveness of the FSCS. Appropriate activities include: conducting interviews with key personnel; observing specific operational procedures; reviewing records; monitoring testing procedures; and sampling materials for analyses.

The checklist may be expanded during the course of the compliance verification, if required, in order to determine compliance to the reference standard. The scope of the compliance verification is established using the plant's compliance history and any risk factors of the product and process.

The compliance verification checklist is an important record of an individual assessment. With each successive compliance history, the previous checklist should be reviewed during the creation of the new checklist.

Opening the compliance verification
The compliance verification begins with an opening meeting with the plant management. At this meeting, the regulatory personnel should explain the scope and objectives of the verification. At this time, industry representatives should pose any questions regarding the compliance verification. Industry management should be encouraged to provide a person to accompany the regulatory personnel during the compliance verification.

When an industry representative accompanies the regulator, a number of benefits are realized:

- Industry witnesses the regulator's observations in real time.
- The industry representative can provide answers to questions immediately.
- The presence of a company representative can facilitate interviews with plant personnel (alternatively, it may hinder interviews).
- The transparency of the regulatory process can be improved, which can also improve communications.
- Industry may benefit in terms of a learning process.

Wrap-up meetings should be held with the processor each day that the compliance verification continues, and the plant management should be informed of the verification progress. Of course, when any issue of critical significance is found (i.e. relating to the safety of a food product) the plant management must be informed immediately.

The investigation

To conduct the compliance verification, the regulator follows the plan established by the compliance verification checklist and proceeds with the investigation. The investigation is a series of planned activities to collect objective evidence in support of the inquiry. The results of all findings are recorded.

Objective evidence is qualitative or quantitative information, records or statements of fact pertaining to the implementation of a quality management program, and is based on observation, measurement or tests. Examples of objective evidence include:

- information contained on company records
- facts related during an interview with a plant employee
- inspector observations
- laboratory results
- product inspection results, and
- measurements made by an inspector.

When a regulator has reason to believe that the FSCS has failed (i.e. it is not being implemented as designed or it is not effective), then objective evidence should be gathered as supporting evidence. An instance of system failure is called a non-conformity. Non-conformities may be either procedural or performance related, and minor or critical with respect to food safety concerns.

Closing the compliance verification

The regulator's findings, including any non-conformities and supporting objective evidence, are documented in the compliance verification report and presented to the company at the closing meeting. Since the regulator has a daily wrap-up meeting with the company, the contents of the final report should not be a surprise to the company. The purpose of the closing meeting is to discuss the findings of the compliance verification with all relevant levels of company management to ensure there is an understanding of the results.

During the closing meeting, the company will be asked to initiate corrective action plans for each non-conformity addressed. When the company has provided corrective action plans (at the meeting or at a later date) these will be reviewed, and if satisfactory, will be accepted by the regulator. The regulator will also verify at a later time that the corrective action is implemented by the company.

4.5.4 Common barriers to regulatory verification

Government agencies may encounter difficulties in moving from a traditional inspection approach to a regulatory verification approach. It is important for regulatory agencies to recognize barriers and develop the appropriate strategies to overcome them. Common challenges to instituting regulatory verification follow.

Cost of change

Over the long term, a regulator verification approach to assessing industry control systems will cost less and provide greater assurance of safety in the food system than a traditional approach. However, the transition period between programs will be a burden on both monetary and human resources. Governments should expect to go through several years of implementation before realizing cost savings. In assessing the cost of introducing a systems approach to food inspection, government should not exclude the cost of failing to move in this direction.

People

In contrast to the standards and pass–fail criteria associated with traditional inspection methods, the regulatory verification approach may be uncomfortable for some regulators. Regulatory personnel may indicate a resistance to change, which can be attributable to moving from known to unknown territory. For example, the movement to regulatory verification is accompanied by the need for new skills, such as audit techniques. Resistance to change may be overcome by communicating to the public, industry, and regulatory personnel the benefits of using a systems approach.

Regulatory personnel may feel uncomfortable with the new knowledge and expertise required when using the industry system verification approach. They will need a good understanding of the hazards associated with food products, food processes, and controls.

Regulatory verification relies on the personal judgement of trained, experienced personnel. The approach recognizes human experience, memory, perception, and cognitive thinking as perhaps the most powerful assessment tools available.

4.6 The Canadian approach

4.6.1 Background

In Canada, at the federal level, responsibility for food safety and inspection has been shared by four federal departments: Agriculture and Agri-food Canada; Health Canada; Industry Canada; and Fisheries and Oceans Canada. To enhance the effectiveness and efficiency of the Canadian food inspection system, the government of Canada amalgamated the food inspection activities of these four departments on 1 April 1997 into the Canadian Food Inspection Agency (CFIA), which reports to the Minister of Agriculture and Agri-food Canada.

The CFIA is responsible for all federal food inspection activities extending from the production of animal feeds, along the food production chain, to distribution and retail stages. The Agency's mission is safe food, animal health and plant protection. For the first time in Canada's long history of food inspection, a single federal agency is now solely responsible.

4.6.2 The current Canadian food inspection environment

The Agency's formation brought together 14 different food, animal and plant health inspection programs. These programs share many similar fundamental food inspection principles, goals and objectives but have evolved independently under different environments and different forces. Consequently, various approaches and methods are used in achieving each program's goals. The level of government intervention and industry responsibility varies from program to program. Products of equal risk, but in different commodity groups, are subject to different inspection regimes.

These programs are not static but continue to evolve to meet the changing demands of the marketplace. Currently, many programs are in transition, moving from a traditional inspection approach to a systems audit approach. In order to prevent continued program divergence and to ensure program evolution follows a common set of principles and a common discipline, the integrated inspection system (IIS) concept was developed.

4.6.3 Traditional inspection approach versus systems approach

Historically, food inspection programs have been based on traditional inspection methods, such as sampling and analysing products for visible indications of disease, spoilage or contamination. Food processing establishments are inspected to assess the basic level of sanitation and hygiene to prevent contamination or spoilage of products. Laboratory analyses are carried out to detect a limited number of chemical, physical and biological indicators of unacceptable products or processes. Although these traditional methods are valid, they are snapshots in time and are reactive to problems already present in the finished food. This approach does not provide consumers with the desired level of confidence in the food products they consume.

The adoption of a systems approach demands that the food industry understand its responsibilities and is knowledgeable of the regulatory and food safety requirements associated with its business. Under a systems approach, industry is required to develop and implement effective control measures to prevent food safety hazards, fraudulent products, diseased animals or potential plant pests from reaching the marketplace. These control measures are based on science and must be validated to ensure they are effective. The system verification approach is preventive and proactive. It moves away from the strategy of "see a problem – fix it" to a "see a cause – prevent it" approach.

In a systems approach, government inspectors have many tools to assess the effectiveness of industry's controls. Inspectors can now review records of production, control measures and corrective actions, etc., gathered by industry over a period of time. In addition, traditional inspection techniques can be used to focus on areas where non-compliance is identified or simply to verify that control measures are effective.

The CFIA has several successful programs that are based on the systems approach:

- *QMP – fish processors.* The CFIA's mandatory Quality Management Program (QMP) applies to all fish and seafood products processed for export from Canada or traded interprovincially. Its objective is to verify that these products are processed under conditions that meet all regulatory, trade and food safety requirements. The QMP facilitates the export of Canadian fish and seafood products by meeting international requirements for HACCP systems. The effectiveness of the QMP program is verified by CFIA inspectors.
- *QMPI – fish importers.* The Quality Management Program for Importers (QMPI) is carried out by importers who have voluntarily chosen to assume additional responsibility under a shared or enhanced QMPI. In these cases, a written QMPI submission must be prepared to address licensing and notification, labelling, ingredients, packaging materials, process controls for canned and ready-to-eat products, storage, final product and recall procedures. The CFIA reviews and approves the QMPI submissions, and conducts verifications of the importer's systems.
- *Food Safety Enhancement Program.* The CFIA's Food Safety Enhancement Program (FSEP) is a voluntary program designed to encourage the development and maintenance of HACCP based systems in federally registered agri-food processing establishments involved in processing dairy products, meat and poultry, processed fruits and vegetables, and shell and processed egg commodities. Under FSEP, the company is responsible for the development, implementation and maintenance of prerequisite programs and HACCP plans. Written document review and on-site verifications are conducted by the CFIA.

4.6.4 The future – the integrated inspection system

Presently, the Agency's inspection programs work well in maintaining Canada's food inspection system as among the best in the world. Naturally, these programs will continue to evolve and improve in order to meet the challenges of new hazards, pests and diseases, and to respond to the advancing globalization of trade. In its first Corporate Business Plan (1997–2000), the CFIA proposed the development of the IIS as the mechanism that will guide the evolution of all Agency inspections programs under a consistent approach.

The IIS can be described as one inspection system for all food commodities, where industry is responsible for controlling its products and processes in compliance with recognized standards and government is responsible for verifying the effectiveness of industry's control systems and making appropriate interventions when necessary.

The objectives of the proposed IIS are as follows:

1. To provide uniform and disciplined inspection strategies for food, animal and plant health that provide the appropriate level of food safety and consumer protection and address the international requirements necessary in order to facilitate market access for Canadian food, animal, plant and forestry products.
2. To provide an effective and efficient food, plant and animal health inspection system, that is open and transparent to all stakeholders.
3. To integrate and interlink the goals, objectives and activities of all players along the food continuum.

Integration under the IIS contains two parts: internal and external.

The internal integration of government inspection programs
The proposed IIS model describes the conceptual framework for the development of the IIS. It contains the IIS reference standard and the IIS verification reference standard. The two reference standards serve as blueprints to guide the development of industry control systems and the government verification system to assess the effectiveness of industry controls.

The reference standard has been based on concepts from the International Standards Organization 9000 series and the fundamental principles of audit. It specifies the quality system requirements designed for application by all stakeholders along the food continuum, including industry, government and third parties.

The reference standard contains 10 basic elements. Its application will be flexible and not all elements will be applicable to all inspection programs. In some cases, the controls may be very basic and focus on maintaining a sanitary environment for handling the food product. In others, there may be comprehensive regulatory and trade requirements and food safety hazards that must be controlled. Inspection programs may be required to incorporate all elements of the reference standard.

The proposed IIS verification reference standard will identify the government's strategy to verify industry control systems. The strategies will be scaled appropriately to reflect industry's control measures. For example, the traditional inspection technique is a strategy used for industries that do not have any control systems in place. Auditing techniques will be used where quality management systems have been implemented. In instances where a third party is involved in the verification of industry's controls, the CFIA will take on a role similar to an ISO registrar to assess the third party's verification system.

External integration along the food continuum

The second component critical to the development of the IIS is the proposed external integration of food safety and quality strategies along the food continuum. The objective is to examine the current food safety control strategies and determine if they are the most effective and appropriate. This will provide the opportunity to interlink and build on the different control measures and address any risks that may have been currently overlooked.

It is proposed that this initiative will be conducted with representation from all segments of the food industry – from production to retail, consumer groups, federal and provincial food inspection agencies, and academia. The task will be to map out the food continuum for their specific products and then, with scientific support, identify the hazards along the food continuum related to safe food, consumer protection and market access. The current control measures will then be evaluated to determine their effectiveness and efficiency in reducing hazards to acceptable levels, preventing them, or eliminating them altogether. In cases where the inspection system can be enhanced, the IIS reference standard may be used to develop new control strategies. The strategies will identify the most effective point for control along the food continuum, the control measures that are to be implemented, who is best suited to deliver the controls, how the controls will be verified and who is best suited to verify them.

The strength of the integrated food control system is the involvement of the stakeholders to work co-operatively to achieve the desired outcome. The result will be a food control system which was developed, verified, communicated and implemented by all stakeholders, with far-reaching benefits. Industry will be able to build valuable partnerships and to implement more efficient systems to ensure that its products meet all applicable food safety, regulatory and trade requirements. Consumers and Canada's international trade partners will have greater confidence that the products will meet their expectations of safety and consumer protection. Regulatory agencies will be able to make more effective use of their resources to direct their activities corresponding to the level of risk.

Part II

Safety and quality

5

Assessing supplier HACCP systems

A retailer's perspective

M. Kane, Food Control Limited, Cambridge (formerly Head of Product Safety, Sainsbury's Supermarkets Limited)

5.1 Introduction

This chapter looks at implementating and assessing of HACCP systems in the UK from the point of view of retailers. It looks at how major UK retailers have encouraged the implementation of HACCP systems by food manufacturers supplying them with products for their network of supermarkets. The major multiple retailers' interest in HACCP can be traced back to the 1970s when they began to develop their own-brand range of food products, which became commercially highly successful. Supermarkets were acutely aware of their product liability exposure with own-brand products, which were mostly co-produced for them by third-party food processors and packers. Commercially, they needed to be able to demonstrate 'due diligence' in their safety procedures in the event of a criminal or civil prosecution against them, even before this became a statutory requirement. These pressures led to an early interest in assessing HACCP systems.

5.2 Retailers and the development of supplier HACCP systems

Supermarkets quickly built up in-house food technology departments during the 1980s to monitor their own-brand food supply lines and deal reactively with those food safety issues that inevitably arose. They first started to apply formal HACCP techniques in the mid-1980s with the publication by Sainsbury of a supplier guidance manual on HACCP based on the original Pillsbury text of the early 1960s. The other major food retailers soon followed, developing their own approaches to HACCP planning and implementation. Suppliers with more than one supermarket customer found themselves visited by technologists from each retailer, often with conflicting technical 'advice' on HACCP issues!

Since then the costs of running large in-house technical departments have become a significant competitive disadvantage. In the late 1990s supermarkets began to encourage suppliers to use third-party auditors, approved by the retailers, to audit their safety and quality systems. Concerns were raised about variations in the approach of different third-party auditors. In response to these concerns, the major UK food retailers have, through the British Retail Consortium (BRC), recently agreed a common minimum standard for food safety and quality audits. This standard provides third-party audit companies with a common basis (using HACCP principles) with which to provide a due diligence defence for retailers. UKAS now accredits third-party audit bodies to this BRC standard for retail food supply.

To start with, the major retailers used the HACCP framework as a management troubleshooting tool, because its methodical and logical approach lent itself to the 'reactive' investigation and solution of those food safety incidents that arose on occasions. However, the disruptive costs of product withdrawals and recalls, plus the costs of adverse publicity, soon pushed the supermarkets into developing the use of HACCP systems in a more proactive and preventive way. They began to encourage suppliers to incorporate HACCP principles into their existing quality management systems, i.e. to predict potential food safety issues and build in preventative controls in advance.

Initially, UK retailers found themselves the pioneers in HACCP development, both encouraging their suppliers to adopt HACCP principles and providing the necessary guidance and technical support to allow them to plan and implement HACCP systems. Supermarkets encouraged suppliers in a number of ways, for example by arguing that the successful development of a HACCP system would help a supplier to gain a recognised quality standard such as BS 5750/ISO 9000, at a time when many companies were seeking such accreditation as a source of competitive advantage, though this has now been superseded by the BRC standard. Supermarkets also provided support for suppliers contemplating developing a HACCP system. Sainsbury's first manual for suppliers on HACCP systems, for example, provided a basic template for HACCP planning and implementation, including guidance in such areas as the following:

- selecting a HACCP team and team leader
- constructing detailed process flow charts as a basis for hazard and CCP identification
- procedures for classifying the severity of hazards and isolating CCPs
- procedures for auditing an implemented HACCP plan.

Supermarket staff also provided expertise, for example, on the range of microbiological hazards that suppliers needed to consider, their conditions of growth and methods of control. Initially, the development of HACCP systems was a learning process for both the retailers and the first suppliers implementing HACCP systems. Some of the problems faced by these suppliers included the following:

- failing to apply the right criteria in hazard and CCP analysis (for instance, by conflating safety and quality issues), resulting in an over-complex and unwieldy initial HACCP design
- unfamiliarity with new systems and responsibilities from line management and staff.

However, supermarkets were able to benefit from their unique position in working with a network of suppliers, able to compare differing approaches to and experiences of HACCP implementation. This experience was used both to strengthen the expertise on which suppliers could draw and develop retailer skills in auditing HACCP systems and making suggestions for improvement. Supermarkets also embarked on a major HACCP training programme for their technologists in the 1990s to consolidate their in-house expertise. Over time, as acceptance and implementation of HACCP principles have become more widespread, supermarkets have increasingly been able to make HACCP implementation a precondition for supplier selection. The technical role of the major retailers has moved towards the administration of a framework of third-party auditing of established supplier HACCP systems.

5.3 Assessing supplier HACCP systems: routine audits

The retailers' experience of HACCP implementation by their suppliers has shown that the most successful have been those with a number of common characteristics:

- well-designed and managed prerequisite systems
- existing quality systems, often with certification to a standard such as those within the ISO 9000 series
- a management culture focused on food safety and continuous improvement.

Companies with these characteristics have the right foundation on which to develop a HACCP system. This includes a basic framework of process monitoring and documentation, which can provide a useful starting point for HACCP planning. More importantly, such companies have the enthusiasm and organisational skills to plan and implement a well thought-out HACCP system. It is perhaps not surprising that supermarkets have developed a policy of selecting suppliers from companies with these characteristics, and that, as a result, HACCP implementation has proved a generally rewarding experience for both retailers and suppliers.

There are a number of ways in which supermarkets assess the effectiveness of a supplier's HACCP system. These include:

- regular supplier audits

- monitoring of adverse customer complaints trends, followed by audit visits to isolate and resolve the problems underlying the trend.

The second of these methods is discussed in more detail in the following section which looks at the ways that retailers conduct trend and cluster analysis, based on customer complaint data. Other indicators include:

- variable product quality noticed in routine monitoring
- media reports.

Routine audits assess the effectiveness of HACCP systems in a number of ways. An auditor would first review the documented HACCP plan. An effective assessment of the quality of the plan requires knowledge of the following:

- the product and production processes
- the typical hazards associated with the relevant raw materials and production processes
- the range of controls that should be in place to monitor key processes
- the CCPs and minimum GMP requirements that should be included in the HACCP plan.

Because they have dedicated food safety expertise within their food technology departments, and the privileged experience of working with a range of suppliers on HACCP planning, supermarket staff are able to bring considerable expertise to this initial inspection of a HACCP plan, benchmarking it against other plans and international best practice. One measure of effectiveness is to check whether the plan is under continuous review, for example by looking for revision numbers and dates. This check can show whether the HACCP plan is a 'live' system, responding to changing circumstances, and being effectively implemented and 'owned' by plant management. Another useful indicator of a 'live' HACCP system is the design of CCP monitoring systems and documentation. Clearly set-out procedures for measurement and recording of CCP data, and clear guidance on remedial action in the case of deviation from permitted values, show the care with which the HACCP plan has been put together and is likely to be understood and implemented by line operatives. Review of the HACCP plan would help to determine the audit schedule agreed with the supplier and any special areas of attention in an audit visit to the plant. Modern electronic process control systems can now provide fully integrated HACCP control with corrective actions highlighted at operator interfaces for all CCPs; with the added benefits of reduced paper with improved audit facility.

In the audit visit itself, there are a number of ways of assessing the effectiveness of the implementation of a HACCP plan. These include:

- the use of scoring systems, based on the audit schedule, to provide a more systematic assessment for feedback of overall performance and particular areas for improvement
- questioning department managers', line supervisors' and operatives' understanding about which CCPs they have responsibility for, how frequently they

are monitored and why, and how they deal with any deviations from the permitted values
- sample inspection of monitoring records for selected CCPs to check that measurements are being carried out in the manner set out, and any deviations acted upon.

Speaking to production line staff is particularly important in assessing how effectively a HACCP plan has been implemented and is 'owned' by the relevant staff in the organisation.

Section 5.5 considers a number of common weaknesses in HACCP planning and implementation. These reflect some of the areas of improvement that experienced HACCP exponents have identified in supplier auditing HACCP systems.

5.4 Non-routine audits: the use of customer complaint data analysis

Gathering objective evidence that a supplier is either in or out of control is a key function of the retailers audit team. Supermarkets are in a unique position to analyse how well their suppliers' HACCP systems work. This is because, as the consumer's first port of call, supermarkets are in the front line to capture customer complaint data and analyse it statistically using computer software.

Two systems of customer complaint analysis have been developed by the leading supermarkets:

1. *Trend analysis*, where quality defects are correlated with point of production and used to target quality improvement initiatives or to delete those suppliers who were unresponsive.
2. *Cluster analysis*, where serious customer complaints alleging food poisoning or product contamination are collated as 'clusters' which are analysed for statistically significant correlations. Where the intrinsic risks of the clustered product and the customer complaint symptoms combine in a manner that is capable of professional interpretation as possible food poisoning, an immediate product withdrawal would be actioned and a HACCP-based on-site investigation initiated.

Trend analysis of customer complaint data has long been established in principle, even if in practice the application of statistical techniques had not been uniformly adopted by all supermarkets. Trend analysis allows the identification of the principal sources of product quality problems, and more importantly of the specific causes responsible for the majority of such complaints. In this as in many areas, the 80/20 rule is found to apply, i.e. 80% of problems arise from 20% of the supplier base. Where significant quality problems persist, suppliers may be de-listed by the retailer in question.

Cluster analysis is a more sophisticated concept whereby serious, alleged food poisoning complaints are rapidly screened over a specific time period to establish whether an unexpected number of such complaints in that time period, or cluster, has occurred. This can be established in advance of any trend in such complaints, allowing for pre-emptive action. It relies upon a professional understanding of the normally arising incidence of complaints, an appreciation of the inherent hazards in any product category and information from customer reports of symptoms of the alleged food poisoning.

Today, all the leading supermarkets have adopted a HACCP-based approach to food safety management where trend and cluster data are key inputs for prioritising the investigative and corrective actions of a more focused team of food technologists either in-house or third-party consultants. Supermarkets have become highly adept at managing crises associated with the common causes of breakdowns in HACCP or other food safety control systems.

5.5 Common weaknesses in HACCP systems

A key element in auditing is the ability to identify areas for improvement in a supplier system. The awareness of likely problems helps the auditor to plan an audit more effectively, identify issues more quickly and suggest appropriate solutions. Experience of monitoring and auditing HACCP systems suggests that there are three main areas of weakness auditors should take account of:

1. The design of the HACCP plan.
2. Failure to maintain the HACCP system.
3. Very occasionally, management neglect of safety as a priority.

Some examples of these sorts of weakness are discussed below.

5.5.1 Design weaknesses: infant food

In the mid-1990s, the CDSC (Communicable Disease Surveillance Centre), a government agency set up by the Department of Health (DoH) to monitor human disease in the UK population, noted an increase (i.e. a cluster) in the number of an isolate of *Salmonella* food poisoning in very young children. A total of 16 cases of this *Salmonella* isolate had been reported for the first six months of a particular year compared with 12 and 7 cases for the whole of the two previous years. An investigation was begun on the basis that the statistical significance of the data was beginning to indicate a serious incident.

By the end of the investigation, the persistence of CDSC and the DoH at the early stage of this incident, when the statistical evidence was weakest, would be totally vindicated. At the start of the incident, however, two cases dated to two or more months previously, and of the 14 remaining, only ten had been subject to case interview. Of these ten, only six were reported as having consumed Brand X infant food, while five were reported as having consumed Brand Y. While these data appeared slim evidence of causal association, the CDSC statisticians also knew that of their 40 planned control interviews (that is interviews of similar families in the affected areas who had not reported with *Salmonella* food poisoning), the first 16 had not consumed Brand X infant food.

Such statistics, which can be typical of this type of food poisoning incident at the early stages, were daunting to all but the CDSC statisticians. Within the week that the investigation had begun, the DoH decided to contact Brand X to present the data and the emerging suspicions of a causal link. Brand X, sensibly, decided to withdraw all the product from sale immediately on the precautionary principle. Following technical debate well into the late evening, the decision to have a public recall was made that same evening and announced on the following day, and an immediate factory investigation was instigated.

A significant proportion of withdrawn product samples were subsequently found to be *Salmonella* positive, though all at a contamination level significantly below the previously accepted infective dose. The CCP was that the product was designed and marketed as an infant food, which implied a target consumer group with a greater susceptibility to *Salmonella* infection, and therefore a 'lower than average' infective dose. The possibility of an elevated susceptibility of infants to lower than average *Salmonella* contamination levels had not been adequately considered in the product formulation or process specification, because it had not been adequately considered in the initial HACCP study of the product, or identified as a CCP.

5.5.2 Design weaknesses: smoked salmon

Traditionally, smoked salmon was effectively preserved for ambient storage with over 15% total salt (on water content) together with heavy smoking involving up to 30% dehydration weight loss. Such a product was shelf stable at ambient temperature, and suitable for postal distribution. Such heavily salted, smoked and dried traditional products have long ceased to be organoleptically acceptable to today's consumer. Smoked salmon today is only very slightly salted and dried (3.5% salt in water), and effectively only flavoured with the smoking process. As a result, smoked salmon today relies upon a controlled distribution through the refrigerated chill chain for its microbiological stability and safety, within a given shelf-life.

When traditional preservation techniques were phased out, it became clear, from a number of customer complaints and queries to many retailers, that some customers were still sending smoked salmon through the post to friends and relatives. It was evident that some consumers were not aware of the significance of changes in processing and their implications for product handling and storage. Most retailers responded by briefing staff at their stores to warn against this practice, and asking manufacturers to include clear advice in food labelling about the unsuitability of mail delivery for this product, and the need for proper chill chain maintenance.

In hindsight, it was clear that revising HACCP plans to incorporate new processing and preservation techniques had failed to take into account food safety implications further down the supply chain and, in particular, the need to educate consumers to abandon traditional practices such as sending the product through the post. HACCP design needs to take account of all stages in the food chain, including the likely response of consumers.

5.5.3 Failure to maintain the HACCP system: ropey milk

The term 'ropey' as applied to milk, beer and bread has come down in the history of the food industry. The term 'ropey' is often used without an awareness of the original food derivation. The 'rope', 'strings' or 'slime' refer to bacterial slime produced by bacilli, that is gelatinous and slimy or sticky in appearance. The problem has been known about for many decades, and has long been solved and relegated to the food science history books. Using cluster analysis an incident was discovered in the early 1990s with some pasteurised milk products where customers were complaining about 'slimy' milk, quickly confirmed as 'ropey' milk.

The first reaction of the management in question when confronted with the cluster analysis data was that, as 'the dairy industry hadn't had a ropey milk problem for 20 years', there must have been some error in identifying the nature of the incident. Three weeks later, and following continuing reports of similar customer complaints of slimy milk, they discovered a grossly contaminated rinse water tank that had been added as a secondary increase in cleaning capacity. The added tank had been configured in such a way that while increasing the total volume of final rinse water available, it also inadvertently created a stagnant volume of water. This stagnant volume of water gradually became contaminated with bacilli, and acted as a contamination source for all the final rinse water used in the dairy.

This incident occurred as a result of a management failure to review the original plant HACCP plan in the aftermath of significant plant changes, i.e. the rinse water tank capacity increase. HACCP plans should always be revised after significant plant changes and additions, and this needs to be assessed with particular care by an auditor.

5.5.4 Management neglect: *Salmonella* food poisoning with snack salami

Despite the requirement for 'Best before' date marking, salami is actually a 'Best after' product, but no such labelling designation exists. The reason is that salami is not, as often incorrectly described, a raw meat product. It is a product made from raw meat, but the raw meat protein is denatured by the chemical action of curing salts and bacterial acid production. During this curing process bacterial action and bacterial acid production combine with the curing salts, over time, to produce a safe and delicious food.

The curing process involved in the production of salami requires careful attention to temperature, rate of acidity production and maturation time and conditions. Safe salami product, free from pathogens such as *Salmonella*, can be guaranteed by diligent professional attention to the conditions of production. A traditional product like salami can, however, sometimes be produced where the original understanding of the basic craft and science has been lost, and methods of production are continued 'as they have always been' for generations. The product is still safe as long as the process remains unchanged, even if the original craft and scientific understanding has been lost.

However, in the 1980s, when a new management team decided to produce a snack salami of finger thick dimensions, they changed the process with the fateful result of a major *Salmonella* poisoning incident. The process for normal (approx. 3 inch diameter) salami production was faithfully reproduced, but the critical point was missed that the surface area to mass ratio was critically different. The new snack salami dried much quicker. This meant that the water activity fell faster, crucial microbial activity was suppressed sooner and acidity development was incomplete. Under these new conditions, *Salmonella* was protected from the hostile effects of the curing salts and normal acid production from bacterial fermentation. *Salmonella* tended to survive under these conditions. To make matters worse, any snack product is passed through the stomach more quickly than a normal full meal. In the intestines, where less acidic conditions apply, *Salmonella* can survive more easily to infect the host consumer. This incident demonstrated neglect due to lack of understanding of the basic food science. The hazards arising from the change in process conditions should have been identified by a thoroughly revised HACCP plan.

5.5.5 Management neglect: *Salmonella* contamination of dried baby milk

Ultimately all food safety failures can philosophically be attributed to management failure, but there are some food safety failures that can be attributed to management's failure to learn the lessons of food safety history. The importance of highlighting this problem is not to engage in witch-hunting, but to elevate the importance of continuous training and professionalism in the management cadre.

During the early 1980s a UK brand of dried baby milk suddenly suffered a *Salmonella* contamination problem. The CDSC was capable of statistically associating an outbreak of *Salmonella* food poisoning among babies with a particular Brand Z of baby milk powder. Interestingly, at the start there were no actual contaminated product samples as evidence of product contamination. Looking for actual product contamination presented the proverbial 'needle in a haystack' dilemma. All the evidence was statistical association of disease with product consumption patterns. A curious point was that the *Salmonella* outbreak was all caused by a single strain of *Salmonella* rather than a cocktail of *Salmonellae* strains that normal contamination patterns would create.

In the event it was discovered that a hairline crack in the stainless steel lining of the spray drier had allowed a single cell of *Salmonella* to leak into the rockwool insulation lining of the spray drier. There it had enjoyed a degree of protection from the heat of processing and the chemical sanitisers during cleaning, and with the abundant nutrients of the milk product, had multiplied rapidly. During cyclical processing and cleaning, the *Salmonellae* had migrated to and fro across the stainless steel lining of the spray drier, intermittently contaminating the dried milk product. *Salmonella* is more resistant to heat under dry conditions, and some survived to contaminate the dried milk and subsequently infect some children.

The actual incidence of contamination was very low, hence the difficulty of finding actual contaminated product and the practical impossibility of controlling this problem by product sampling alone, but the epidemiological evidence of *Salmonella* poisoning in the baby population was indisputable. All this was relatively easily determined after the event by judicious professional investigation, using HACCP as an investigative tool, but the question remained as to why the circumstances had not been predicted before the event. The simple answer was that the HACCP studies on the process before the incident had been inadequate. If the whole sequence of events in the incident had never occurred before, the HACCP study would have been limited by previous experience and the failure to predict the problem could have been reasonably accepted. This was not the case, however. The precise sequence of events had occurred in Australia some four years previously. The management in question had failed to keep themselves up to date with events in their industry and apply the lessons from similar incidents. Effective auditing requires an understanding of incidents such as these and their implications for effective HACCP design and operation in preventing their reoccurrence.

5.6 The future development of HACCP

Retailers have identified a number of ways in which HACCP design and implementation can fail. These include:

- quite commonly, a failure in the design of the HACCP plan, e.g. in identifying risk levels for particular customer groups; or ineffective hazard analysis because of a failure to keep up to date with current research and best practice
- quite commonly, a failure to audit the HACCP plan satisfactorily and particularly the impact of process changes on the hazards originally identified in the plan
- rarely, a failure of managers to understand the basic food science, processes and hazards sufficiently to undertake a proper hazard analysis, which can only be dealt with by a recognition of the need for training with the appropriate commitment and resources to acquire the requisite knowledge and skills
- very rarely, a failure to identify customer safety as the primary management responsibility, which can only be resolved by appropriate training and changes in management culture.

Given the increasing prominence of food safety issues, these last two failures are now rare, although they have by no means been eliminated as competitive pressures in the food industry intensify. Indeed, they may remain significant problems as global sourcing brings new suppliers into the food chain who may be less familiar with food safety issues and HACCP systems. For those more familiar with HACCP systems, the kind of design and verification problems outlined earlier are likely to continue to be applied as the pace of product innovation intensifies and new hazards emerge. Allied to these problems will be the need to keep established HACCP systems 'fresh', for example in keeping staff motivated. A programme of ongoing staff training will be important here.

The following discussion looks at a number of possible improvements to HACCP design and implementation, from ways of improving current HACCP systems to how the scope of HACCP itself can be extended.

5.6.1 Improving HACCP analysis: improved process flow diagram construction

Constructing an accurate process flow diagram is the critical starting point in a HACCP analysis. In any process, one of the most likely points that a food safety hazard will occur is where an unplanned process delay or interruption happens. Such a process delay point is often the location where build-up of microbiological contamination can occur.

A process flow diagram system, which aids the identification of any likely delay points, is significantly more efficient as a management tool for identifying hazard points in the process. The original process flow diagram method of the American Society of Mechanical Engineers (ASME) still has much to recommend it in this respect. As can be seen from Fig. 5.1, the ASME method allows a detailed analysis of a particular process against seven key criteria which help identify potential problem areas such as process delay points.

5.6.2 Extending the scope of HACCP: criminal malicious product contamination

HACCP systems have sometimes been seen as having a primarily microbiological focus, confined to particular products and processes. However, HACCP principles can be more broadly applied to other aspects of food safety, for example malicious product contamination. Incidents of deliberate malicious product contamination are now, regrettably, to be regarded as an established criminal practice and food industry hazard. All food manufacturers must acknowledge this and therefore plan appropriate countermeasures. 'Appropriate countermeasures' simply means elevating the protection of product integrity to the same status as all other food safety control measures that comprise normal good manufacturing practice (GMP).

HACCP principles can be employed and specifically focused on this issue. The experience of managing a technical investigation into a criminal malicious product contamination incident that has occurred within the food chain can be

°C Temp.	Time	○ Operation	⇨ Transport	D Delay	□ Inspection	▽ Storage	Item No.	
					●		1	Raw material inspection – to specification
		●					2	Weigh out ingredients into guisti – 750 kg batch
		●					3	Add water to batch size
		●					4	Mix
95		●					5	Heat – Steam jacket
		●			●		6	pH adjustment to 4.0 with lemon juice
	20 mins			●			7	Hold
<40	30 mins	●					8	Cooling
			●				9	Transfer to filling head
		●					10	Fill to sterile bulk tank
12	4 days	●					11	Cooling
					●		12	Final product QC check – pH/aw/Micro
AMB	5 days					●	13	Storage
					●		14	CO_2 gas analysis – QC check
AMB						●	15	Storage awaiting dispatch
AMB			●				16	Transport to Yogurt production unit
<10		●					17	Chill prior to use
					●		18	Check CO_2 gas in headspace & pH
		●					19	Connect to conserve metering pump

Fig. 5.1 Example of ASME process flow chart (hazelnut yogurt conserve).

employed to develop measures to prevent a contamination problem in the first place.

Every food processing plant is a unique facility that will inevitably have its own unique security problems. But every processing plant can be reviewed logically using HACCP principles, and sensible measures can be implemented to improve security without creating the fortress conditions that would interfere with the proper functioning of the plant. Every criminal who contemplates deliberate malicious product contamination must have the motive, the means and the opportunity to carry out a crime. HACCP principles can be employed to assist food processors to frustrate potential criminal product contaminators by identifying those with potential motives, restricting the means available and reducing the opportunities to perpetrate their crimes.

Major retailers now issue guidelines to suppliers for dealing with this problem covering such issues as the following:

- staff training in awareness of the problem, identifying potential motives for malicious contamination (e.g. as a result of certain disciplinary actions)
- using a HACCP approach to identify critical points where the product is unavoidably exposed to potential malicious contamination

- security measures (e.g. the use of closed circuit television (CCTV) systems to monitor high risk areas, control of personnel access to high risk areas, colour coding of staff clothing to identify bona fide workers easily in high risk areas, sensors and other instrumentation designed to detect evidence of contamination, tamper-evident packaging)
- crisis management procedures and robust product traceability.

This approach mirrors that used by retailers to deal with malicious contamination in their stores.

5.6.3 Improving the evaluation of HACCP systems: post-launch monitoring of new food products

When new food products are launched, the degree of pre-launch evaluation is limited to panel testing that rarely amounts to more than a few thousand people. More usually only a few hundred people are involved. These numbers bear little statistical significance to the tens of millions of customers that each of the major supermarkets service each week. The customer complaints monitoring that major supermarkets use is also capable of monitoring customer reactions to new products. In the future such existing systems of complaint monitoring could include specific post-launch monitoring of new foods, looking more closely for any emerging food safety problems, for example nutritional issues among more vulnerable sections of the population such as young children or elderly consumers. In fact post-launch monitoring will come to be recognised as an essential element of 'due diligence'. Genetically modified foods could also be monitored by such statistically-based systems. Data from these exercises could be shared with manufacturers of the products concerned, and any general implications for food safety made public so that it can inform future HACCP analysis. In this way HACCP systems can be more effectively evaluated and improved for the future.

5.6.4 Integrated HACCP control systems

The application of HACCP to basic food processing operations calls for the application of specified control procedures at all CCPs, ingredients tracking systems, monitoring and reporting of key factory conditions, cleaning procedures, batch tracking systems and final product traceability. The ease with which these basic requirements can be integrated by modern electronic process control technology has been under appreciated by food manufacturing management, who still currently rely on paper-based control systems.

Of the seven principles of HACCP, four directly address 'process control' aspects. Specifically, these cover:

1. defining the CCPs
2. setting control limits for each CCP

3. monitoring to ensure each CCP is under control
4. taking corrective action when a CCP is out of control.

By definition, a CCP is therefore crying out for the application of automation technology. Where CCPs are controlled automatically, using the full armoury of electronics technologies such as sensing, vision systems, motion control, temperature/process control, operator interfaces, networking, information management and knowledge management, then the production line will benefit from consistent operation, continuous (fatigueless) monitoring, and the guarantee that these points, procedures or operational steps will be maintained in control within specified limits. And this means safer food production systems, records of non-conformances and corrective actions, full traceability, networked management business process information, vastly reduced paperwork, active operator interfaces for input and instructions, with the final bonus of integrated knowledge management systems based on multi factorial statistical process analysis and diagnostics. The future of HACCP implementation will be in integrated active process control systems using the components of electronic automated process control such as PLCs, touch screen operator interfaces, sensors of many descriptions, PID controllers, distributed controller networks, fieldbus networks, vision systems and RF tagging.

5.7 Conclusions

HACCP is essentially a structured way of thinking about the management of food safety. Its effectiveness depends on the ability of food producers to make the most of the analytical framework it provides and, in particular, the knowledge and skills of HACCP teams in such key areas as hazard analysis and CCP identification. Businesses often fear that the main problem with a HACCP system is the administrative burden they think it will impose. In fact the greater challenge it presents is in analysing the way food products are manufactured and the resulting hazards in a systematic way which some businesses have never previously employed.

In reality, because the quality of HACCP teams varies, the quality of HACCP planning is variable and, in some cases, poor. Retailers auditing supplier HACCP systems also continue to discover within some businesses a lack of understanding of HACCP principles among HACCP teams. The effectiveness of implementation also varies with, in some cases, a lack of 'ownership' of the HACCP system by operational staff. Alongside some of the weaknesses identified earlier, common problems include:

- failing to deal with prerequisite systems first as a foundation for HACCP analysis
- over-complex and unmanageable HACCP plans that confuse hazards and quality issues and identify too many control points as 'critical'
- lack of understanding among CCP monitors of their role, purpose and

importance, resulting in poor monitoring and record keeping, and failure to take corrective action.

These problems impose a particular responsibility on the quality of auditing in identifying weaknesses, developing understanding and encouraging initial HACCP systems to continue to improve until they reach a satisfactory standard. Government and industry bodies also have a role to play in spreading understanding of HACCP principles and offering support and advice in such areas as the training of HACCP leaders and HACCP teams and information on hazards. It will also be important to strengthen the perceived commercial benefits of implementing HACCP systems. One development here is the emergence of new business insurance products which recognise and reward HACCP implementation. Technology will also play a part: improvements in automated process control, for example, will help to make CCP monitoring easier, and developments in real-time, on-line analysis of key hazards will help to validate and improve the design and operation of HACCP systems. Cumulatively, these forces can make the potential of HACCP systems to manage food safety effectively throughout the food industry into a reality that will benefit consumer and industry alike.

In all these areas, auditing, both by businesses themselves and by retailers and others, is the driving force for improvement. If planned and executed effectively, it provides a way of systematically testing the robustness of a HACCP system. A key issue for auditors is that auditing should be a dynamic and iterative process, building on the experience of previous audits and informed by relevant industry and subject experience. In this respect retailers, with their network of suppliers, have a valuable role to play, refining the auditing process in the light of their cumulative experience of the strengths and weaknesses of supplier HACCP systems.

6

TQM systems

D. J. Rose, Campden and Chorleywood Food Research Association, Chipping Campden

6.1 Introduction

Organisations looking to develop their business operations through the current volatile economic climates need to establish clear objectives as to how the various elements of the business need to perform to ensure continuing growth and viability. In order to achieve these objectives it is further imperative to have mechanisms in place to monitor performance and also to provide a process by which change can be implemented in those areas of activity which need strengthening. Total Quality Management (TQM) is a management tool which provides that opportunity.

In its broadest sense TQM provides a business system by which the whole organisation can be harnessed to meet the needs of customer requirements. It is important to emphasise that TQM is not merely a technical standard but encompasses both the technical and business operations. The fundamental requirement for a successful TQM system is to have good management practices, TQM alone cannot provide this and any systems implemented will only ever be as successful as the staff involved.

The purpose of this chapter is to describe the key elements that need to be considered when setting up a Total Quality Management system. It provides information on the typical range of quality systems that may already be in place within an organisation and looks at how these can be used to bring together all of the requirements necessary to achieve a TQM system. It further explains the key steps necessary to begin development of the system and the implementation process required. Finally the key monitoring processes needed to confirm successful implementation and for continued improvement and development of the system are explained.

6.1.1 Defining quality

Despite the preponderance of quality assurance texts, quality standards, and definitions of quality, many people are still confused by the term quality. In the early incarnations of quality management, quality assurance and quality control were often used synonymously. However the importance of differentiating between reactive quality management (quality control) and pro-active quality management (quality assurance) was quickly realised. More importantly the benefits to be derived from the wide ranging implications of quality assurance were soon realised and capitalised upon by practitioners. The concept of TQM takes the now more familiar quality assurance requirements, as exemplified by BS 5750/ISO9000, one step further and seeks to view ALL operations and processes that a company utilises as being inherently important to their overall business performance and quality of service parameters.

According to BS 7850, Total Quality Management may be defined as follows – 'Management philosophy and company practices that aim to harness the human and material resources of an organisation in the most effective way to achieve the objectives of the organisation.' On a slightly different tack, Margaret Thatcher once paraphrased quality very succinctly, 'The combinations of features in a product which ensures that customers come back for a product which does not.'

However, it is important to realise that the objectives of the organisation can be multifaceted and reflect other primary business needs as well as the more obvious product quality issues. TQM systems should therefore be capable of incorporating objectives as diverse as customer satisfaction, business growth, profit maximisation, market leadership, environmental concerns, health and safety issues and reflect the company's position and role within the local community. One over-riding principle must be for the TQM system to ensure compatibility with the needs of current legislation in all its guises – food safety, business practices, environmental and waste, employment rights and health and safety.

The need to meet the ever-increasing demands of customers for improved reliability and quality of product have fuelled the need to consider TQM systems. Supplying 'just-in-time' manufactured products with short shelf-lives to the retail outlet in a reliable and dependable manner, pressure on margins to provide cheap yet wholesome foods, and the continuing need to provide evidence of safe food production have all added to the requirement to consider the totality of the chilled food business operation.

Unfortunately for staff tasked with considering TQM systems there has been much confusing literature produced on the subject. Various titles have been used to describe TQM systems, e.g. Continuous Quality Improvement, Total Quality, Total Business Management, Company Wide Business Management, Cost Effective Quality Management, Integrated Management Systems. Suffice it to say that the objectives of the various schemes have all been synonymous and I refer the reader back to the definition of TQM given earlier from BS 7850. The challenge to practitioners of TQM is usually not with the title given to the system, but rather to understanding their business well enough to identify all of the key elements required to be set up and managed within the umbrella of TQM.

This analysis of the key business processes may be achieved by a variety of different means. Most critical to the analysis is the ability to collect suitable and usable data which reflects the process. The use of data collection forms, performance data, market research, productivity information or financial data may all be appropriate. Analysis of the data to extract useful and usable outputs may be performed by a variety of different techniques. BS 7850 recommends affinity diagrams, brainstorming sessions, cause and effect diagrams, flow charts and tree diagrams to analyse non-numerical data. Control charts, histograms, Pareto diagrams and scatter diagrams may be useful for numerical data. By understanding all of its business processes companies are able to define the process, implement controls, monitor performance and measure improvements. This is the fundamental basis of Total Quality Management.

6.1.2 Quality assurance systems

The foundation for any quality system is to be found in the fundamental principles of Good Manufacturing Practice (GMP). There are many GMP guidelines available for the manufacture, handling and preparation of various kinds of product such as, for example, chilled foods (Department of Health, 1989; Institute of Food Science and Technology, 1990; Chilled Food Association, 1995, and 1997; National Cold Store Federation, 1989). All focus on the key technical requirements for safe, hygienic, good manufacturing practices, allied to good storage, handling and distribution practices. In this context, these can be considered the fundamental technical objectives or standards to be achieved. Currently a large number of targeted quality assurance systems have found favour throughout the food industry.

The most prevalent of the formal quality systems is still the BS EN ISO 9000 (BS 5750) suite of standards incorporating specifically BS EN ISO 9002 for production facilities and BS EN ISO 9001 for production operations incorporating new product development activities. ISO 9000 systems provide the advantage of laying down formal management controls for production activities, but also can easily be extended to other critical business activities such as purchasing, sales and distribution. Many operations have already extended their ISO 9000 systems into a TQM system by encompassing their other critical business processes.

Manufacturing production sites have now also been forced to consider the requirements necessary to meet the requirements of Hazard Analysis Critical Control Point (HACCP) systems based on *Codex Alimentarius* definitions. As well as providing the key control measures necessary to understand the mechanics of producing safe food, HACCP systems also provide the basis around which to build production control systems and to ensure product quality in the operation.

Documentation of HACCP plans to meet the seven fundamental principles of HACCP as laid out by *Codex Alimentarius* is also required. These documented plans, together with associated operational procedures, records of operation and evidence of maintenance of the critical control points, often form enough of a basis for production activities to be controlled and managed by using the HACCP plans as a quality system – see principle 7: 'Establish documentation concerning all procedures and records appropriate to these principles and their application.'

More recently businesses have needed to consider the impact of their operations on the environment. Moves to standardise environmental control and management have been formalised within ISO 14001. This international standard 'Environmental Management Systems – Specification for Guidance and Use' has strong links to ISO 9001 and covers issues such as policy statements, process control, system structure, training, awareness and competence, system documentation, checking and corrective action, preventive action, record keeping, system auditing and management review. The stated aim of ISO 14001 is to 'provide organisations with the elements of an effective environmental management system which can be integrated with other management requirements'. This approach is an obvious lead in to the concept of incorporating environmental objectives within a TQM system.

Yet more recently, safety systems have been targeted for incorporation within the suite of quality system functions and BS 8800 ('Guide to Occupational Health and Safety Management Systems') provides a framework within which to manage safety systems and safety training activities. Given the increasing importance of staff occupational safety and the need to minimise exposure to potential litigation, manufacturers are well advised to treat this area of activity seriously. Companies may also have an interest in other systems related to staff training – i.e. the Investors in People standard within the UK, organised through local Training and Enterprise Councils, which requires proper evidence of structured training programmes for staff, records of all training activities and clear benefits being derived from both staff and employers from their training programme.

In a critical key development the British Retail Consortium (BRC) has now issued its core Technical Standard for Companies Supplying Retailer Branded Food Products. This standard is being used by a large number of UK retailers as the definitive standard for suppliers and terms of business are being agreed which include the requirement for companies to meet this standard. The BRC standard itself focuses on a large number of essential and recommended good manufacturing practices and is underpinned by the need to establish supporting management systems to back up these manufacturing practices. In essence six key areas are involved, HACCP systems, quality management systems, factory environmental standards, product control, process control and personnel. Implementation of the standard is being handled through third-party inspection bodies whose remit is to ensure compliance of the operating site with the standard. In some cases, as for the European Food Safety Inspection Service

(EFSIS), the inspection bodies have incorporated the BRC standard within their own inspection standard to provide an even more rigorous examination of the operating site.

All of the quality systems mentioned above have essential core elements and similarities. Most importantly the critical elements of control can easily be related to the core business functions of the site. These, taken hand in hand with the key legal, safety and environmental control measures all sites are obliged to adopt, offer a comprehensive and complete set of frameworks within which to develop a total quality management system.

6.2 The scope of a quality system

This section summarises the essential business processes to be considered when addressing Total Quality Management systems concerned with the production of chilled foods. The next section deals with the necessary administrative detail of developing the quality system.

6.2.1 Raw materials, purchasing and control

- Raw and packaging materials should be purchased to agreed specifications, and from suppliers capable of achieving those specifications. Suppliers should be audited for quality and safety.
- Raw materials should be stored under hygienic conditions that prevent contamination by microorganisms, insects and other pests.
- Stock control systems should be used for minimising storage times. Coding systems should be used to ensure traceability.
- Inspection of raw and packaging materials should complement the suppliers' quality systems.
- Control and release should be under the responsibility of a competent technical person.
- Non-conforming raw materials should be recorded and investigated to identify and rectify problems.

6.2.2 Process control

- The HACCP approach should be used to identify critical control points as part of developing process specifications and to determine monitoring programmes.
- The HACCP plan must be suitably verified and the control points demonstrated to be sufficient to control the product.
- The arrangement of plant should minimise the likelihood of cross-contamination.
- Plant should be hygienically designed.
- Plant cleaning schedules should be developed and implemented.

- Critical measures such as time, temperature and quantity must be recorded throughout the production process.
- Sampling regimes must be set up to measure product quality and safety throughout the production process.
- Contingency plans need to be in place to cover any possible major safety issues that may arise.

6.2.3 Premises

- Premises should be constructed to minimise the risk of contamination.
- Premises should be maintained to a hygienic standard commensurate with the degree of risk.
- Where used, high care areas must be suitably constructed and all necessary control measures for their operation implemented.
- Suitable waste disposal facilities need to be in place.

6.2.4 Quality control

- Clearly defined product specifications and quality standards should be used to supplement HACCP analysis in identifying non-safety quality issues.
- Product quality (in terms of sensory characteristics) should be defined to meet the specifications given above, and agreed with clients.
- Product quality should be verified to ensure acceptability before release and on-going monitoring checks should be in place to prevent major defects arising.

6.2.5 Personnel

- Personnel should be trained in hygienic practices and other quality requirements of the job.
- High standards of personal hygiene are essential.
- Clothing appropriate to the task is required.
- Appropriate sanitary facilities are required.
- Medical screening is required.

6.2.6 Final product

- Inspection must take place to determine conformance with the product specification and freedom from any foreign body contamination.
- A system for isolating non-conforming product is required.
- The type and level of inspection should be determined from HACCP.
- Critical testing and inspection should be done by competent laboratories.
- Where technically important, or for legal reasons, checks on packaging should be done.
- Records of inspection must be kept.
- Shelf-life validation is required.

- A system for monitoring complaint trends is required.
- Product release should be by positive approval.

6.2.7 Distribution

- Arrangements must be made to maintain product integrity in the distribution chain.
- The level of batch traceability must be commensurate with the risk of recall.
- A recall system should be developed and tested.

The above list is by no means comprehensive, but indicates the breadth of considerations to be addressed in chilled food operations. The task is complex and requires a high degree of skilled management. It should be developed and implemented as part of cohesive quality system.

6.3 Developing a quality system

Developing a quality management system to meet the requirements of your business is a complex task. Not only do the elements described above need to be considered, but also such factors as management responsibility, documentation and auditing. The standard model for quality systems for some years now has been the ISO 9000 series of standards, the international standards for quality systems. The general applicability of ISO 9000 to the food industry has been demonstrated by its successful application in many production facilities. However ISO 9000 has suffered from criticism over the years due to its unfriendly nature and the perception that it does not lead to quality improvement, only control and standardisation of processes. Consequently quality management, as exemplified by ISO 9000, has often been seen as being on a parallel track to business management, and not as an all embracing TQM system.

Changes now introduced within the revised ISO 9001: 2000 standard help address this failing by focusing the system back towards quality improvements, process development, continuous improvements and customer satisfaction. The above comments notwithstanding, the fundamentals of ISO 9000 still provide the basis around which to start the development of the quality system, a TQM system being built by encompassing all of the other business process requirements onto this model.

6.3.1 Management responsibility

The importance of senior management commitment to the quality system cannot be over-emphasised. If quality is established as a board room priority, all other parts of the organisation will follow and become intimately involved in the process. Similarly, by defining key responsibilities for all levels of activity, those staff whose actions can influence the quality of the food or the process under which it is manufactured can be identified and made aware of their responsibilities. This is so that errors do not occur through it not being clear who is responsible for various actions, for example, who monitors a chill room temperature, or who should carry out a particular quality control test.

Senior management must review the continuing effectiveness of the quality system at periodic intervals. Key information sources would include data from internal audits, non-conforming product records, quality control records on conformance to specifications, and customer complaints trends.

The second key role of the management review process is to establish mechanism for improvements and new initiatives. The evaluation of key data such as quality markers, which can be used to measure client satisfaction, and performance measures, which can be used to measure the efficiency of the delivery process, should be encouraged. Based on the analysis of these data, together with the data mentioned above, senior management can identify business processes which would benefit from improvement or re-design.

6.3.2 Documentation of the system

Effective documentation of the procedures and actions required to achieve the required quality is an essential part of the quality system. Such documentation can be used for reference and for training purposes. It reduces the risk of misunderstandings arising from oral communication. All documents should also be controlled so that personnel do not work from obsolete documents. There must be a means of circulating new procedures and withdrawing superseded ones, and a controlled means of making changes to procedures. Staff discipline with documentation also needs to be instilled so that only the current versions of documents are used.

6.3.3. Customer requirements

A clear understanding of customer requirements is essential for any business operation. Mechanisms to ensure that sufficient information is captured from clients prior to manufacturing, need to be set up. These will include fundamentals such as manufacturing details, supplier arrangements, product specifications, delivery times, quantities and packaging. However consideration must also be given to other matters such as legal requirements, environmental, employee and safety considerations.

6.3.4 Raw material control and supplier quality assurance

The quality of raw materials and the suitability of packaging materials has a considerable effect on the final quality of chilled foods. However, relationships with suppliers go well beyond these criteria and extend to the quality of service, prices and financial stability of the supplier. These factors must be combined together to achieve a smooth and profitable relationship between vendor and purchaser. The objective must be to build a confident partnership between the two so that the purchaser can rely on the vendor as much as on 'in-house' departments.

There are a number of stages to go through in achieving this objective. It should be noted that all purchased materials which can affect product or service quality should be included in this programme. Often capital and services items (e.g. motors, pumps etc.) are omitted, and faults are only detected after installation. Clearly this does have an impact on the overall efficiency and quality of the operation and should be included.

Supplier quality policy

There should be a stated and preferably written policy. This usually takes the form of a summary of the principles involved:

- mutual co-operation; the partnership
- prior contractual understanding; agreeing specifications
- methods of evaluation
  - audit
  - inspection at source
  - inspection on receipt (the policy is to accept only material meeting the requirements)
- plans for settling disputes
- feedback on performance
- vendor responsible for delivery to standard.

Contractual understanding

There is little point in trying to develop a partnership with suppliers unless there is a clear understanding about the objectives to be achieved. This usually takes the form of a contract covering for example, material specifications, delivery parameters, responsibilities for quality including those for verification, access to supplier, procedures for settling disputes. It is important that all these parameters are agreed and verified prior to signing the contract and entering a supplier onto an approved list.

List of approved suppliers

The purchaser should maintain a list of approved suppliers. Lying behind this should be a set of procedures which describe the stages of approval. There are two main criteria to be considered here:

1. financial capability and stability
2. ability to meet specification.

The latter can be assessed in a number of ways:

- auditing supplier's quality system
- vendor's previous performance

- vendor's reputation
- tests on representative samples.

Auditing suppliers
The objective here is to establish the supplier's ability to meet agreed requirements. Auditors should be trained to conduct this activity promptly and efficiently. The auditors observe the manufacturing facilities, buildings environment, plant, quality procedures and implementation of such procedures. Other evidence to collect includes management attitudes, workforce attitudes, quality control records and so on. Often auditors will also look at financial and technological aspects.

Feedback on performance
It is absolutely essential in building the partnership that the vendor receives prompt and accurate feedback on performance.

Performance data can be collected from a number of sources.

- *Raw material conformance:* sources of information here include the vendor's own inspection records, incoming inspection records. Most non-conformances in this area are clear and are well 'flagged' because usually they result in a delay in deliveries or production.
- *Process conformance:* non-conformances here are less easily detected but at least should be reviewed during audit. It may be written into contractual requirements that process non-conformances are communicated to the purchaser.
- *Procedural non-conformance:* similar comments appertain here as for process non-conformance.
- *Raw material unfit for use:* this is the worst scenario where a non-conformance is not detected until it fails either on the production line or in distribution or in use (complaints). The impact is usually severe, affecting ability to sell the final product. Despite the severity of the problem, it is often difficult to gather sufficient evidence to inform the vendor of the fault.

Feedback should be given on a regular basis so that each non-conformance is not seen by the vendor as a 'complaint'. The main message here is to transmit good as well as bad news. Where possible, evidence should be incontrovertible. The best evidence is records and samples. Regular meetings with suppliers will ensure that the positive feedback is given. This helps to support the partnership when exceptional communication of non-conformances is necessary.

6.3.5 Process control
All aspects of the production of chilled foods having a direct bearing on the quality of the final product must be specified, documented and recorded to ensure that failures due to inadequate control are eliminated. Critical control point monitoring as identified by HACCP forms part of this requirement. Action

when results are outside specifications must be clearly identified with responsibility allocated. The HACCP principle should be used throughout the production process and include raw materials and final storage and distribution. It can be used for all potential hazards including inadequate quality as well as safety.

HACCP includes the assessment of potential hazards, prescribes for the elimination of available hazards and sets tolerances for the hazards that cannot be eliminated in the processing of a food. It defines the appropriate control measures, the frequency of their application, the sampling programme, the specific tests to be applied and the criteria for product acceptance. Since HACCP is an ongoing dynamic process, analyses will need to be reviewed in the light of new hazards and changes in the process parameters. HACCP has the potential to identify areas of control where failure has not yet been experienced, making it particularly useful for new operations.

The following definitions are used in HACCP:

- Hazard analysis is the identification of potentially hazardous ingredients, storage conditions, packaging, critical process points and relevant human factors which may affect product safety or quality.
- Critical control points (CCP) are the processing factors of which loss of control would result in an unacceptable food safety or quality risk.

Carrying out a HACCP analysis

To carry out a HACCP analysis, a formalised and structured approach is needed. A broad base of information is required and will therefore require specialist knowledge from many disciplines, since safety and quality assurance cannot be categorised by a single discipline. The first stage of an analysis is to obtain a detailed flow diagram for the process under consideration, including methods and schedules of production, preparation and transport of raw materials. Many of the considerations will be influenced by issues specific to the factory.

The second stage of an analysis identifies the essential characteristics of the product and its use, enabling definitive conclusions to be drawn about the hazards or potential risks which will threaten either the consumer or the product. Consideration is given to food storage conditions, formulation of the product, the packaging used, the expected customer handling practices and the target consumer group.

The third part of an analysis is consideration of all the stages in the process, taking into account realistic process deviations. Critical stages in the process are identified which must be controlled adequately to assure safety – the critical control points (CCP). A judgement of risk must be made using one of three basic methods: probabilistic, comparative or pragmatic. The choice of method depends upon circumstances and the basis for any judgement should be recorded. Such judgements require a high degree of expertise and experience and should only be made by suitably qualified people. Ideally, the opinion of more than one 'expert' should be sought. If process details are incomplete, the most unfavourable assumptions must be made unless, for example, there is a long, proven history of the raw materials presenting no hazard to the process or the product. The final stage of an analysis is to devise standards for and effective procedures to monitor critical control points and appropriate corrective action as mentioned earlier.

Monitoring of critical control points
Monitoring of CCP may be best accomplished through the use of physical, microbiological and chemical tests, visual observations and sensory evaluations. Monitoring procedures, including those which take the form of a visual inspection only and do not involve measurements, should be recorded on suitable checklists. These checklists should show details of the location of the CCP, the monitoring procedures, the frequency of monitoring and satisfactory compliance criteria. For chilled foods, the cleanliness of equipment is a CCP. Therefore a hygiene maintenance schedule must be devised that specifies what should be cleaned, how it should be cleaned, when it should be cleaned and who should clean it.

When monitoring of CCP takes the form of inspection, particular attention should be given to temperatures of food, hygienic practices and techniques of handling foods by workers, whether employees are ill or have infections which can be transmitted to the food and opportunities for cross-contamination from raw to cooked foods. Control options also include arrangement of plant to minimise cross contamination, building maintenance and cleaning, and staff training.

6.3.6 Inspection and testing

From HACCP, a schedule of testing for raw materials, intermediate and final products is developed. Methods of tests must be defined, responsibility for testing and the acceptance criteria drawn from appropriate specifications. At each stage, product should not be released until inspection is complete. If release takes place earlier, a traceability system must exist for recall purposes. The time required to complete microbiological tests on chilled product is problematical here. However, most microbiological tests are used to monitor the success of process control rather than for testing product characteristics. Untested, tested, approved or rejected materials need to be clearly marked to avoid any possibility of confusion.

All test equipment used to demonstrate compliance with a defined specification or to control a critical process should be of known accuracy. Required measurements should be identified, the measuring equipment calibrated at defined intervals, against acceptable physical or nationally recognised standard references. Calibration methods should be described and adhered to, and the calibrated equipment must be identified as such. Records of calibration should be kept, and if a calibrated instrument is found to be inaccurate, then a designated person must review the situation and decide what action should be taken in respect of materials previously measured with that instrument. Where necessary, critical tests should be performed by a suitably accredited laboratory, either in-house or external.

Any product which is found to be outside specification should be segregated to prevent inadvertent use. The product should then be destroyed, re-worked, or re-graded. In exceptional circumstances, customers may be prepared to accept the product, but not if safety is in question. Re-work must be controlled strictly. Causes of non-conforming product should be identified and action taken to prevent recurrence. Complaint trends should also be monitored and corrective action taken as appropriate.

6.3.7 Handling, storage, packaging, delivery

This is extremely important for chilled foods. Precautions must be taken to protect product quality throughout production and the chill chain. Hygiene precautions, including vehicles and chill storage, pest control and restrictions on access would be included here. The legislative requirements for 'food handlers' and legal constraints on labelling, date coding and food contact materials should be addressed here. The means of temperature control, monitoring and recording are critical. Determination and control of shelf-life through stock rotation must be included.

Decisions on the extent of, and method of, traceability must be reached with respect to the risks of recall. A fully documented and workable recall system must be implemented. The system should extend to distribution centres, the trade and in extreme cases, consumers. The recall plan must be tested to ensure its effectiveness.

6.3.8 Records

An effective record system is essential. The control of records, including their identification, safe storage, retrieval and disposal, should be defined. It should be clear when records can be disposed of, and who is responsible. The most important records are those which demonstrate that what has been specified has, in fact, been achieved. These include process control and inspection records. However, in order to satisfy the legislative requirement for 'due diligence', other evidence will be required such as records of internal audits, management review, supplier audits, HACCP records, temperatures in distribution, corrective action, cleaning and training.

6.3.9 Quality audits

A scheduled system of internal quality audits is essential to ensure that all procedures are implemented and working effectively and that instructions are written down and followed. These audits are an effective management tool for monitoring the success of the quality system and ensuring that everyone is working to the system.

System audits should be undertaken by trained personnel independent of the area being audited. Audits are carried out by a process of observation, interview and examination of records. Any non-compliances should be recorded and referred to the responsible manager for timely corrective action. A follow-up to ensure that the action has taken place should be carried out, and records of the action kept. The results of audits should be reviewed by senior management. The audit schedule should cover all aspects of the quality system and include compliance with legislative requirements and voluntary Codes of Practice.

6.3.10 Training

All staff must be trained to fulfil their responsibilities with regard to tasks undertaken which affect quality. Training needs should be reviewed, the needs identified should be fulfilled and records kept. Staff education and training is often a most useful option for control of hazards such as microbial contamination. In addition to hygiene training, there is another special training requirement for the food industry to be considered here: training for sensory analysis. An attempt to ensure that judgement of product quality in this respect is objective must be made.

6.4 Implementation

Implementation of any quality initiative is difficult. Change, often perceived to be change for changes sake, is not always acceptable to staff. It is therefore imperative that the correct empowerment is given to the implementation and that it is introduced and explained to staff in the right manner.

6.4.1 Chief executive commitment

The ramifications of a TQM strategy are too large for them to be considered at anything other than the highest levels within a company. Ideally the idea to implement the system should come from the chief executives themselves. Alternatively it may come from other sources. Whatever the source it is not worth starting the exercise until the right level of approval and commitment is achieved. Once the senior management are on board with the idea they must throw their whole weight behind the initiative, any perceived weaknesses will be exploited by opponents of the scheme.

6.4.2 Steering group

As a first step, set up a steering group to manage the implementation programme. This group should consist of staff drawn from each of the principal areas of operation in the company: this should include sales and marketing, purchasing, production, distribution, technical and finance. The group should be headed by a member of staff with sufficient managerial experience and should be accountable to the board or directly to the chief executive. The steering group should also appoint the person to be designated as the 'management representative', whoever named, who will be responsible for the maintenance and control of the system in the future. This person is concerned with ensuring that documentation proceeds smoothly and that documentation is controlled. He/she is also responsible for the internal audit system. This so-called 'Quality Manager' may have other duties within the quality system, but this should not be to the detriment of 'ownership' of the system by all the constituent parts.

6.4.3 Initial status

The steering group should arrange for the two key activities to be carried out: (a) a definition of current business processes carried out in all parts of the business and (b) based on this information define the scope within which you wish to implement the TQM system. This is a key decision and must be based on a sound understanding not only of the essential processes which support delivery of your products to your customers, but must also understand all of the support functions which help to maintain that delivery mechanism (e.g. finance, maintenance).

Once the scope of activity is defined it is imperative to carry out an exercise to establish the level of benefit that may be obtained from the introduction of the TQM system. Typically this can best be achieved by carrying out a quality costing exercise. Quality costing will determine the operational costs of not doing things right, such as wastage in manufacturing, loss due to non-conforming product, down-time on equipment. Based on the findings of this exercise it is possible to estimate the possible company wide benefit of introducing the quality system. These potential savings can then be reviewed by senior management, and a firm commitment to establish the system made. If necessary the scope of the TQM can be reviewed at this point to ensure that the areas covered will lead to maximum return.

6.4.4 Planning

The steering group should draw up the implementation sequence and agree timescales with all appropriate parties so that a plan can be made. The plan must cover all elements of the implementation including process analysis, documentation, implementation, training and PR elements. The group should monitor the plan of implementation. If there are problems to overcome in achieving the plan, the steering group must be sufficiently senior to overcome blockages. If the plan cannot be achieved for unavoidable reasons, the steering group must give an account of this to the chief executive.

6.4.5 Quality policy

It is important that the chief executive writes a quality policy for the company. This can be anything from a relatively simple statement to something more complex. At its simplest it states that the company is dedicated to meeting customer requirements. If it is the intention to work to a recognised quality system, then a statement to the effect that the company wishes to comply with the requirement of, for example, ISO 9002 should also be included. However, other statements about business culture and objectives can also be included. For example, employee welfare, environmental policies, position in market place and so on, can all be considered. It is best not to make the policy too lengthy or complex.

6.4.6 Briefing

It is the steering group's responsibility to initiate and co-ordinate briefings throughout the project. During the initial phases, this would be an announcement (from the chief executive) about TQM or ISO 9000 explaining what it is and the reasons for pursuing the course of action. Such briefing should be to all employees, but would be more detailed for some depending on the level of involvement envisaged. Also the substance of the briefing will depend on the seniority of the audience.

It is best to keep the briefings short and to the point initially; more detailed training can follow later. A big 'launch' package with trumpets blaring is not the best course. Small and informed focusing on the facts, the importance of the initiative and not underestimating the amount of work involved will get the message home. The seniority of the person carrying out the briefing speaks volumes about the importance of the mission.

6.4.7 Structure of the quality system

It is essential that the structure of the quality system is agreed at an early stage. This is best embodied in the documentation. There should be three levels to the documentation, although for a small company this may be kept under one cover. The three levels are:

1. policies
2. procedures
3. work instructions.

Policies should be used to state the company's intent with respect to key elements of the system, e.g. policies with respect to purchasing arrangements or staff training.

Procedures will form the bulk of the system and will provide the detailed instructions as to how principal operations are carried out. These form the bible from which the company will be expected to operate and will be judged (audited) against.

Work Instructions provide the 'shop-floor' level of instruction needed by staff. These should be formulated so that anyone coming to a job for the first time can, with a small amount of training, carry out the job effectively. Examples of work instructions could be how to make out an invoice, a purchase order or a customer order. Other examples include those in production for the basic operation of a machine or on how to carry out weight checks.

All levels of the manual may need to be supported by record forms and these should either be incorporated with procedures/work instructions, or clearly separated off and identified. Setting the structure of the system and document numbering and cross-referencing will save a lot of trouble and retracing later. It is clear that documenting the quality system is a major task and it needs to be thought through clearly. Experience also teaches that the best systems are those written by the staff actually involved in the task being described. This simple device also ensures a wide level of personal involvement with the development of the quality system and helps to provide ownership throughout.

6.4.8 Quality manual(s)

The essential parts of this need to be written in draft form at an early stage because it sets in writing the structure referred to above. It usually contains the policy and headline procedures covering each key area of the quality system. It is usually used for overall guidance and should be available to customers. Therefore it should not contain anything of an overly sensitive or confidential nature. Each operational function within the company must agree to the contents of the quality manual as it applies to them. Better still, to gain ownership, if they actually write those parts which apply to them. The organisation and management authority must be clearly defined at this stage.

6.4.9 Quality improvement

Once the plan has proceeded far enough to ensure a reasonable understanding of the current business processes, the key task of planning for quality improvements can begin. The mechanisms for achieving this are varied and will need to be tailored to suit individual circumstances. Most critically it is essential to be able to measure the process, either in terms of inputs, outputs or throughputs. BS 7850 (ISO 9004-4) deals with a variety of techniques used for quality improvement and these will need to be considered. It should be remembered that improvements may need long-term solutions and that the implications to other elements of the business must be considered. However, failure to maintain momentum in this area will impact significantly on the usefulness of your TQM system.

6.4.10 Staff training

The system designed will be of no use whatsoever unless sufficient time and resource is allocated to training and educating staff in the requirements of the new system. Therefore sufficient time must be allocated and planned into staff training. To meet both new skills training and new working practices, but also any implications that the new TQM ethos may bring to the business, i.e. the need to participate in quality circles, ability to generate improvement suggestions and the need for all staff to be committed to the ideals of the system.

6.4.11 Launch

The system can be launched when it is felt that all key elements are in place. These do not have to include all of the proposed quality improvements. Remember the system is designed to be a continually changing system and evolution to new procedures and practices should be a natural progression. Staff should not be concerned if parts of the system are not perfect, again improvements will be identified as the system evolves. It is much more important that the system accurately reflects the current business processes. Often the benefit of TQM comes with time when the quality system is used to collate data and information about the performance of the business and these are used to target and develop improvements. Once launched internal mechanisms to monitor and control changes to the system should be made operational as well (e.g. internal auditing systems, document control and document change systems).

6.5 Performance measuring and auditing

As indicated above, once the system is launched it is imperative to measure performance and to seek quality improvements. In fact, the most powerful part of any quality system is its ability to measure performance and gain improvement through corrective action. There are a number of tools available to do this within the quality system.

6.5.1 The internal audit system

Regularly measuring compliance with the stated system is a powerful means of determining its effectiveness. The corrective action that ensues from an audit ensures that the system is kept fresh and up to date, reflecting the latest requirements of the company.

6.5.2 The external audit system

Once formal certification and approval of the system is sought, external auditors for the inspection body will visit regularly to ensure on-going compliance with the relevant standards. Again useful information and corrective actions can be obtained from these visits. It is also possible that key customers may wish to audit the systems to give themselves confidence in your ability to provide safe wholesome food. These visits combine the benefits of an external inspection of the system with the specific requirements that the customer may have, enabling these requirements to be built into the overall operation.

6.5.3 Non-conforming product

Any such incidents must be investigated thoroughly. The reasons for non-conformance must be investigated and corrections made to the quality system and operating processes to prevent recurrence.

6.5.4 Conformance to specification

It is necessary not only to ensure that all product conforms to its final specification, but also to determine proximity to target of all measured parameters – both product and process. Clearly this serves two functions: to detect deteriorating trends early, and to detect persistent divergence from target while still within specification. Both may lead to corrective action.

6.5.5 Customer complaints

These should be treated like non-conforming product. They should be investigated thoroughly. Any deleterious trends must lead to corrective action.

6.5.6 Quality action initiatives

One of the key ingredients of any quality system, often referred to unkindly as the weakest link, is an organisation's own staff. However, staff also provide a company's greatest innovative resource. Involvement of all staff with the development of new ideas, process discussion groups and quality circles will enhance process efficiency, detect poor performance and lead to corrective action.

6.5.7 Performance measures

By setting performance measures for each key business area, or delivery process mechanism, the efficiency and performance of each key area can be monitored. Improvements, as well as declining performance, can be tracked and all elements of the business put into a measurable framework. Typical performance measures can extend beyond simple productivity related issues and may incorporate elements as diverse as: energy monitoring, waste management, sales lead successes, new product development time and customer satisfaction surveys. All these sources of information on performance should be subjected to senior management review. It is essential that senior management have the opportunity to review and take appropriate action at regular intervals.

In addition, feedback of performance to staff is an essential means of motivating staff to further improvement. It is quite easy for staff to be involved in performance measurement but not see a clear picture, because they see individual results rather than trends. Such feedback can be in the form of summaries of audits (based for example on a points system), trend graphs for conformance to specification or information on customer complaints.

6.6 Benefits

The achievement of total quality management or a good quality system is a never-ending road to improvement. Those who have embarked on this journey have found a number of benefits.

6.6.1 Economic

Generally, the operation is more cost-effective. This is achieved by getting it 'right first time'. There is a reduction in the amount of wasted material; productivity is increased as a result of the orderliness created. There is a reduction in the number of customer complaints. Machine efficiency improves and manufacturing capacity increases.

6.6.2 Marketing

By meeting customer needs consistently, there is an opportunity to secure the customer base, and to build sales success. Customers are more confident in the consistency of product and they see a commitment to quality.

6.6.3 Internal

A number of benefits are achieved within the operation. Staff morale improves because staff know what is expected of them. There is increased awareness of quality and a commitment to quality. Communication improves and staff are better trained. There is much improved management control with greater confidence in the operation, a reduced amount of 'fire-fighting', a uniformity of approach to procedures and a mechanism for continuous improvement.

6.6.4 Fulfilling legislative requirements

A good-quality system is of great benefit in demonstrating that attention has been paid to complying with legislative requirements, particularly those of due diligence. The quality system provides documented evidence of its functioning through written procedures, of its success through the records, and of its ability to improve through audits and review.

6.7 Future trends

A key change is nearly upon us at this point in time, the BS EN ISO 9000 series of standards have been revised and the new ISO 9001 (2000) version was published in 2000. The new standard marks a significant change to many areas of the old ISO 9000 standard and addresses many of the key criticisms of the old standard. In particular there is a change to using the process model approach so that individual businesses can suit the application of the standard to their own operations, rather than have the 20 key system elements imposed on them. Secondly the emphasis of the new standard will be firmly routed in the principles of continual improvement and meeting customer requirements. To this end specific requirements will be needed to measure and ascertain performance of the business with respect to quality and the ability to meet its customers' demands and requirements, howsoever defined. Finally the new standard includes the requirement to effectively communicate with customers and to manage all relevant streams of information passing through an organisation. In summary the essence of the new standard should help to ensure that you keep your existing customers by focusing on their needs, rather than the internal needs of the business.

It can also be predicted for the future that the involvement of customers, and particularly retailers, in the production supply chain will continue to grow. Requirements on production facilities to ensure that their products meet the needs of retailers is therefore imperative and the moves towards synergistic business relationships between suppliers and retailers should be encouraged. The continuing use of third party schemes to audit and assess production premises will obviously continue and the current standards being applied will develop with time. The challenge to all involved in this process is to ensure that the inspection standards are demanding but technically achievable to ensure safe and reliable food production.

Given the higher publicity now given to any food safety problem by the media, it is an inevitable consequence that governments will react to the media attention by raising standards through increased legislative input. In the UK the launch of the new Food Standards Agency (FSA) will enhance this process. It will be interesting to note whether the FSA will develop a highly prescriptive approach to safety matters or whether it will continue to place the emphasis of control onto manufacturers themselves.

Finally, the environmental pressures being placed on the whole of society will impact on chilled food businesses like all others. TQM systems which seek to address environmental issues as well as production management issues should be applauded and encouraged. The European Union is keen to progress the ideas of an Integrated Product Policy (IPP) within all manufacturing areas. This approach takes a more holistic total life-cycle approach to products and looks into the total environmental effect on all elements of production, packaging, delivery. This ensures that the environmental impacts of the individual elements are acceptable and that any changes made do not create savings in one area by passing on the impact to another area. Backing IPP up by focused environmental audits and certified product labelling will ensure that consumer marketing can also be focused on this key area.

6.8 References

BRITISH RETAIL CONSORTIUM, (2000) *Technical Standard and Protocol for Companies Supplying Retailer Branded Food Products*, BRC, London.

CHILLED FOOD ASSOCIATION, (1995) *Class A (high risk) area best hygienic practice guidelines*, CFA, London.

CHILLED FOOD ASSOCIATION, (1997) *Guidelines for good hygienic practice in the manufacture of chilled foods*, 3rd edn. CFA, London.

DEPARTMENT OF HEALTH, (1989) *Chilled and Frozen. Guidelines on cook-chill and cook-freeze catering systems*, HMSO, London.

FAO CODEX ALIMENTARIUS COMMISSION, (1997) *Food Hygiene Basic Texts. HACCP Principles*, FAO/WHO, Rome.

INSTITUTE OF FOOD SCIENCE AND TECHNOLOGY, (UK) (1990) *Guidelines for the handling of chilled foods*, 2nd edn. IFST, London.

INTERNATIONAL COMMISSION ON MICROBIOLOGICAL SPECIFICATIONS FOR FOODS (ICMSF), (1988) *Micro-organisms in Foods. 4: Application of the hazard analysis critical control point (HACCP) system to ensure microbiological safety and quality*, Blackwell Scientific, Oxford.

NATIONAL COLD STORAGE FEDERATION, (1989) *Guidelines for handling and distribution of chilled foods*, NCSF, London.

7

Auditing HACCP-based quality systems

N. Khandke, Unilever Research, Sharnbrook

7.1 Introduction

The food industry today is increasingly under pressure from the outside world. Food legislation is becoming more comprehensive internationally, standards are tightening and inspection authorities are better trained and have a greater understanding of the hazards and their means of control. Similarly, consumers are more aware, have higher expectations and are concerned about food safety and quality, and the media is quick to pick up food-related stories. In order to remain competitive in the marketplace, meat product companies are changing their approach to product safety and quality. They are moving away from systems based on checking the finished product, to a system of assuring safety and quality through design and control of manufacturing and supply chain operations.

In order to facilitate this change in the trading environment, food producers are adopting standardised systems, or frameworks, within which quality systems can be developed and demonstrated to customers and regulatory authorities. The two major systems currently utilised to manage quality systems are Hazard Analysis Critical Control Point (HACCP) and the ISO 9000 series of quality standards. The ISO 9000 system describes 20 elements required to build a quality system (not all elements are required for each of the different standards) (see Table 7.1). The basic premise of the system is that the producer defines systems and procedures, developing a quality system for his whole operation, documents these procedures and demonstrates compliance with his own internal standards. Because of its structured nature the ISO 9000 system offers the added benefit that certification can be gained from third party certifying bodies, to demonstrate to customers that you have a documented quality system in place.

Table 7.1 The 20 elements comprising the ISO 9000 standard

Element	ISO 9001	ISO 9002	ISO 9003
Management Responsibility	✓	✓	✓
Quality System	✓	✓	✓
Contract Review	✓	✓	
Design Control	✓		
Document Control	✓	✓	✓
Purchasing	✓	✓	
Purchaser Supplied Product	✓	✓	
Product Identification and Traceability	✓	✓	✓
Process Control	✓	✓	
Inspection and Testing	✓	✓	✓
Inspection and Test Measuring Equipment	✓	✓	✓
Inspection and Test Status	✓	✓	✓
Control of Non-Conforming Product	✓	✓	✓
Corrective Action	✓	✓	
Handling, Storage, Packaging and Delivery	✓	✓	✓
Quality Records	✓	✓	✓
Internal Quality Audits	✓	✓	
Training	✓	✓	✓
Servicing	✓		
Statistical Techniques	✓	✓	✓

HACCP is a different tool for identifying and controlling product safety hazards, and unlike ISO 9000 is specific to a line and product. HACCP is internationally accepted and is mandatory in many countries. External normalisation companies and agencies are beginning to offer certification services for HACCP, but this is still in its early stages. However, it is likely, through pressures from customers, that HACCP certification will become more of an issue in the future.

A fundamental process within any quality system is auditing. Auditing is not a new concept, but in the past may often have been viewed as a tool for 'checking up' on a company, or policing the company's systems. Auditing is in fact the main tool for driving continuous improvement, by identifying weaknesses in a quality system and recommended changes for improvement. The two major types of audit applicable to safety and quality are the technical audit and the system audit and these will be addressed in this chapter. A common misconception is that anybody can turn up to a company with a blank sheet of paper and audit the company. This is definitely not the case. Audits must be carefully structured and planned, and must be carried out by trained personnel. The major areas of auditing discussed in this chapter are:

- Scope
- Standards
- Preparation
- Format
- Assessment and scoring

- Follow-up
- Frequency

By carefully addressing each of the above areas, a company can develop comprehensive, effective auditing systems for both internal auditing of their own quality systems, and external auditing of suppliers and third party producers.

7.2 HACCP and quality systems

The majority of processors in the meat industry now accept the fact that the traditional approach of testing a product to detect defects, post production, is statistically unsound, gives no assurance that defective or hazardous product is not released onto the market and provides no opportunity for remedial action.[1,2] As a result, many processors have moved away from this traditional 'quality control' approach to more preventative systems based on design and operational control. In order to facilitate this change producers are adopting standard quality systems, such as the ISO 9000 series,[3–7] HACCP[8] and Total Quality Management (TQM)[9] to name but a few. All the above quality systems share a common element, in that they do not provide a company with a ready-made quality system, but define a framework upon which a company can build quality management systems of the required complexity and focus to enable the consistent manufacture of products of a defined quality. The ISO 9000 quality management series offers the additional facility in that the systems and procedures making up the system are formally recorded so that they can be assessed externally and accreditation/certification given if the system meets the requirements of the standards.

There is extensive information in the literature on the quality systems mentioned above, and it would be futile to try to cover all the topics here. However, we should briefly consider the main systems currently favoured.

7.2.1 The ISO 9000 series

The ISO 9000 series of standards for quality systems[3–7] were published in 1987 and were based upon the British Standard BS 5750[10] and a similar Canadian standard.[1] The ISO 9000 system is comprised of five separate standards. ISO 9000 'Quality Management and Quality Assurance Standards – Guidelines for Selection and Use', and ISO 9004 'Quality Management and Quality System Elements – Guidelines', offer advice and guidance on selecting the appropriate standard and implementing the guidelines. The standards themselves are encompassed in ISO 9001–9003. ISO 9003 covers the quality system for final inspection and test and is not normally applicable to food processors. ISO 9002 covers the quality system for production and installation and is the standard most commonly sought in the food industry, and ISO 9001 is the quality system for design/development, production, installation and servicing, and is the most comprehensive of the three standards.

The ISO 9000 standard is composed of 20 requirements (see Table 7.1) which guide a company into the areas which need to be contained within the quality system. Not all the requirements are relevant for all the standards (ISO 9001 uses more than ISO 9003). The standard itself does not define specific criteria for any of the 20 requirements, but the standards do give guidance on what is required in each. It is up to the company to define the specific criteria required in each section.

There are a number of key features which need to be mentioned with regard to ISO 9000.

1. The ISO 9000 system, as a quality system, normally specifies a quality system for the whole company, covering all quality-related activities.
2. The ISO 9000 quality system is based on the contracts and relationships between customers and suppliers.[1]
3. The ISO 9000 system requires companies to define their own standards, systems and procedures, which they believe will result in the production of product of a consistent quality.
4. In order to gain certification in ISO 9001–9003, the company only needs to define their own standards, to document these standards and associated systems and procedures, and to demonstrate to the assessor that they adhere to these internal systems. There is therefore always the chance with ISO 9000 that a company will not have covered all critical elements for product quality or safety within their internal standards, but nevertheless may achieve certification by demonstrating compliance with those set.

The method by which ISO certification is achieved varies, depending on which certifying body is used, but in general the certification process involves:

- Selection of the appropriate standard and the development of internal standards, systems and procedures covering the 20 elements
- Pre-review of documentation by the third party certification body to identify any early non-conformances, and subsequent remedial action. In a labour-intensive industry this may result in considerable training
- Formal assessment, in house, by the third party certifying agency
- Correction of any non-compliance
- Certification
- Maintenance and reassessment (normally six-monthly maintenance visits and a full review every three years). (This may vary depending on the certifying body used.)

7.2.2 HACCP

Although it is probably fair to say that HACCP predates ISO 9000 (and the BS 5750 series before this), it was not until the publication of HACCP in its current form, based on the seven principles, in the late 1990s[8] that HACCP has come to the fore as a key safety system utilised in the food industry.

The various features of the HACCP system have already been discussed in this book. However, we can draw a number of comparisons with the ISO 9000 system. The first point to mention is that HACCP is a quality management system, and is similar to ISO 9000 in that it provides a framework on which a system can be built. HACCP does not come 'ready made' and a company implementing HACCP will establish criteria to control hazards, based around the requirements defined in the standards.[8] The HACCP system, however, does have a number of important features, distinct from the ISO 9000 system.

1. HACCP as a quality system focuses on product safety, and is targeted at individual production lines and products. This is unlike ISO 9000 which specifies a quality system for the whole company.
2. Although HACCP provides an empty framework, the safety hazards, limits and in many cases the controls for many of the food processes are very often universally accepted and quantified. This makes it easier for a company to gain information on the hazards and controls relevant to a particular food process. It also has the effect of making it easier for an inspector to assess the completeness and technical accuracy of a HACCP plan.

7.2.3 Total Quality Management (TQM)

TQM is unlike HACCP and ISO 9000 in that it does not provide a rigid framework within which to build up a system. TQM focuses on continuous improvement, through the participation of employees in identifying and implementing improvements, and focuses on 'delighting the customer'. TQM therefore provides a philosophy, culture and discipline within which quality systems such as HACCP and ISO 9000 can be built and operated.[11]

7.3 Establishing benchmarks for auditing

Auditing is a fundamental part of a food safety or quality system, whether it be auditing to certify a supplier or a quality system (such as seen in the ISO 9000 system), or internal auditing to assess compliance to Good Manufacturing Practice (GMP), to verify a HACCP plan or to monitor internal compliance to quality systems and procedures. An audit can be defined as a 'systematic evaluation of a system against a set of defined criteria'. Audits are often viewed as being surreptitious checks on companies' systems, with the auditors being viewed as policemen. This should not be the case, and if an audit is perceived in this way it is not being carried out correctly. An audit is a quality tool which allows an auditor to assess performance against a set of criteria. The main purpose of an audit is to drive continuous improvement by identifying areas of weakness which may pose a risk to product quality or safety (and hence a business). There are essentially two types of audit, each of which can be further subdivided into a number of types of audit. At the broadest level audits can be defined as:

- Technical audits
- Systems audits.

Technical audits are generally of a limited scope and performed by technical experts in a specific field, such as microbiological safety, hygienic design, or thermal processing. This type of audit will examine a particular process in detail to assess its technical performance against set criteria. In most cases the criteria set for such audits will be defined externally, in either national legislation or industry codes of practice. The technical audit is more often used to assure the manufacturer that the products manufactured, and the processes or unit operations employed, meet a minimum set of requirements to ensure the safety of the end product.

Systems audits are more commonly applied in the food industry and are not necessarily carried out by technical experts. A systems audit is examining compliance with a set of systems or procedures which make up a company's quality system. The systems or procedures covering supply or production procedures may be internationally or nationally defined, but in most cases will be developed internally by the company. The most commonly recognised systems audits in the food industry are those of the ISO 9000 certification system. The key issue with regard to systems audits is that, where the systems are developed internally, they do not necessarily ensure the quality or safety of the product or process. The absence of a 'judgemental element' can be a problem with the ISO 9000 system where the approach of 'say what you do, do what you say, show that you have done it' can get a company certified as ISO 9000 without the company addressing the critical safety or quality issues within the product or process design.

Within the two audit types above, companies will be carrying out, or receiving, audits of different types, the main being:

- Internal audits
- External audits
- Regulatory audits
- Certification audits.

These types of audit will be discussed later in this chapter.

7.3.1 Establishing the ground rules for an audit

Irrespective of the type of audit that will be carried out, there are a number of ground rules which must be followed to ensure that the output of the audit can be used for reporting and improvement. No matter how experienced the auditor, auditing is not simply a case of turning up to the company or department to be audited with a pen and paper to see what you can find. When this approach is used, it inevitably leads to omissions and inconsistencies in the audit process and

Table 7.2 Main elements required in setting up a successful audit system

Element	Rationale
1. Scope	Defines the type and limit of the audit
2. Standards	Define the depth of the audit
3. Preparation	Allows the auditor to develop an understanding of the product, process and standards
4. Format	Determines the method of the audit, e.g. using check lists, questionnaires
5. Assessment and scoring	Describes the method by which the audit will be evaluated
6. Follow-up	Checks progress against an agreed action plan resulting from an audit
7. Frequency	Defines how often audits will take place

assessment. The key elements of an audit, which must be considered, are shown in Table 7.2.

Scope

The scope of an audit is determined by a number of factors, the two most important being the type of audit being undertaken, and the resources available to carry out the audit. The scope of the audit will be made up of a number of different elements. The first element is whether the audit is a technical or systems audit, together with the type of audit (internal, external, etc.). This level will immediately determine the type of auditor required to carry out the audit, as a technical audit will require specialist expertise in the subject area being audited.

The second element should define what the audit will cover. This is always an important question and is more often than not determined by the resources available. HACCP audits will, by the nature of the HACCP study, be product and process line specific. ISO 9000 audits focus on the company's quality system as a whole. The common trap is to focus on in-house operations during the audit, which may result in critical elements which are important for product quality and safety, but which lie outside the core manufacturing process (upstream or downstream from the processing establishment), being missed. As a minimum, a company's quality audit system should include upstream audits as far as the raw material supplier or primary producer (e.g. farmer). These audits should cover how the supplier manages their own upstream and downstream supply chain, but it is often impracticable actually to audit these elements yourself, and downstream audits extend as far as the end user of the product (for retail goods this would normally be down to the retail outlet) or in the case of a further processing plant the inwards goods reception.

Standards

All audits should be carried out against defined standards. Without standards there is no benchmark or frame of reference, and the auditor's personal belief becomes important in defining what is acceptable and unacceptable. Audits of this nature are rarely satisfactory and can lead to disagreements between the auditor and the company or department being audited over the action points raised. Another consequence of not setting fixed or published auditing standards is that it becomes almost impossible to draw conclusions when trying to evaluate the results of different audits, especially where different auditors are used, because the audits will have been carried out to different standards.

The standards used will depend on the type of audit. For any audit, local legislative requirements which may be agreed with the local veterinary service will be important, but in many cases a company's internal standards may well be stricter than the local legislation. For internal audits, internal procedures and specifications form the basis of the standards against which the audit is carried out. These internal standards should include any published GMPs, and should cover the control, monitoring and corrective actions defined in the HACCP plan. For external audits, e.g. supplier or third party producers, it is more normal to use external standards or industry guidelines. Two good examples are the 'General Principles of Food Hygiene' produced by the Codex Alimentarius[12] or the 'Food and Drink Good Manufacturing Practice Guidelines' produced by the Institute of Food Science and Technology.[13] These are two of many such guidelines which can be useful. When carrying out an external audit, it is important that the auditor takes note of any internal standards being applied by the third party, specifically those defined within the HACCP plan, to assess how well the company is adhering to their own standards.

In all cases the standards to which the audit is being carried out, and its scope, should be mutually agreed in advance of the audit (it is not the objective of the audit to 'catch people out').

Preparation

The key to any successful audit is preparation, whether it is an internal ISO 9000 audit of a department, a HACCP audit of a line or a complex audit of an external supplier. Auditors should familiarise themselves with the scope of the audit and the applicable standards well in advance of the audit. For internal audits they will need to familiarise themselves with the process, products, systems and procedures being audited (it is not good practice to allow auditors to audit within their own department of the plant, and it is good practice to rotate auditors within a company to avoid auditors becoming over familiar with any area or department).[14]

For external audits, the auditor may not be familiar with the product or process in operation because in the meat industry processing covers the scope of processes from slaughter and butchering right through to the preparation of cooked, sliced meats. It is therefore important that the auditor is pre-armed with knowledge of the following:

- The typical hazards associated with such processes and materials
- The controls which should be in place

- The limits within which the process should be capable of working
- The minimum CCPs which should be included in the HACCP plan and GMP requirements
- Product usage
- The process stages and personnel involved, etc.

Format
The audit format determines the method of the audit. There are many different approaches to auditing, each having their own benefits and shortcomings. Whatever approach is used, it should be designed to aid the auditor in covering all the areas defined in the scope of the audit. Some of the more common approaches are:

- Experience based
- Check sheets
- Questionnaires.

Audits based only on experience should generally be avoided, due to possible inconsistencies and the difficulty in interpreting their results. This type of expert audit is more suited to technical audits which are carried out by technical experts and have a very narrow scope. The outcome of this type of audit will be a technical evaluation of a line or process.

Check sheets are the simplest form of 'organised' audit. They normally consist of a series of simple questions designed to cover specific elements of a process or quality system, together with a set of check boxes for each question which can indicate 'Yes' or 'No' at the simplest level to an indication of 'fully compliant', 'partially compliant' or 'non-compliant' in more complex cases. Check sheets often have scores allocated to the individual questions to allow an overall score to be calculated. Scoring is discussed in more detail later in this chapter. Check sheets can be very useful for internal auditing, especially hygiene and GMP auditing, and their relative simplicity enables them to be used by less experienced auditors. The nature of a check sheet is that it is very regimented and guides the auditor in specific directions. This type of audit is less likely to look at areas outside the checklist which in certain situations may provide relevant data for the audit. For example, a check sheet may look at the temperature of a meat slicing operation, it may check that the slicer is clean and that the records of cleaning and disinfection are adequate. However, an auditor using a check sheet is unlikely to pick up whether the slicer is hygienically designed or being operated correctly. Check sheets are therefore more suited to operations where the technical evaluation of suitability has already been performed and the auditor is required to check that systems in place are being adequately performed (verification). Check sheets are therefore particularly suited to the regular auditing of a defined set of specified activities, such as internal hygiene auditing, verification of HACCP systems and ISO 9000 type internal audits.

Check sheets are not well suited to auditing unfamiliar premises (third parties) as their scope is too limited. However, it is often very useful to develop standard check sheets which can be sent ahead of the audit, with the request that they are completed and returned to the auditor before the audit. These can then be very useful in making an initial assessment as they can often identify areas where attention needs to be focused during the audit.

Audit questionnaires come in many different guises and are widely used for auditing. Audit questionnaires differ from check sheets in that they ask open-ended questions which are a prompt for the auditor to cover a specific (subject) area of the processes or systems in a plant, rather than the specific yes/no type of questions used in a check sheet. Effective use of the open-ended style of audit questionnaires require that auditors are experienced in the topic of the audit and must understand the requirements set out in any standards that are available. A good auditor will use each question in the questionnaire as a starting point for a discussion in a particular subject area with the personnel involved, and will not move on to the next question until they have assured themselves that the personnel involved understand their role in processing and that the company being audited is, or is not, complying with the requirements. When preparing a questionnaire, care must be taken that the questions guide the auditor into all relevant areas, but also that they give the auditor enough freedom to fully investigate issues in sufficient depth. This is illustrated below where we ask the auditor to look at the same subject, traceability (i.e. the ability to trace a particular material from its origin to the retail trade or consumer), but in different ways.

1. Does the company (plant) have a lot traceability system in place?
2. To what extent can a company (plant) trace products in the marketplace?
3. Lot identification on packs, bins or product is an essential tool for product recall and helps effective stock rotation. Each container (primary pack) of food should be permanently marked to identify the producer and lot.[15]

Question 1 is very restrictive and more suited to a check sheet. It leads the auditor to make a yes/no assessment and relies on the experience of the auditor to actually go beyond the simple issue of whether a traceability system is present to look at its suitability and extent.

Question 2 is more balanced and asks the auditor to look into traceability to determine whether a system exists and whether or not it is suitable. This question requires that the auditor knows what the applicable standard or internal requirement for traceability is, and is able to judge the level of compliance.

Question 3 is not in fact a question but a quote from the standard on which the audit is being based. This serves two purposes. It firstly tells the auditor to look into traceability during the audit. However, because the question is a quote from the standard, it also tells the auditor what is required. It is important to note that this does not mean that the auditor need not prepare, or be familiar with, the standards. It does, however, provide a convenient *aide mémoire* for the auditor to use during the audit.

Question 1 is not suitable for use in an audit questionnaire, and it is advised to use the approach given in question 2 or 3 above when developing audit questionnaires.

Assessment and scoring

The information collected by all audits needs to be evaluated. The methods by which the evaluation is done are very dependent on the type of audit carried out. The audit process will generate data which informs an auditor how well the activity in question complies with the given criteria defined in the standards. Criteria have been mentioned several times, but it is at this stage that they become very important. When making recommendations, based on non-compliance to a standard or criterion, these must be based on non-compliance with the agreed criteria, such as a temperature, stock rotation regime or hygiene standard. It is not good practice for the auditor to make recommendations based on personal belief, as these will be open to debate. A non-compliance based on an agreed standard, whether it be an internal standard such as a work procedure, or an internationally agreed standard, is much more likely to be agreed and accepted by the company or department being audited.

There are no fixed rules determining the amount of information handed over to the plant being audited at the end of the audit. For third party or supplier audits, it is common only to give an indication of 'Pass' or 'Fail', rather than a detailed written report. It should be remembered that one of the main purposes of auditing is to drive continuous improvement. The auditor should therefore leave an agreed list of recommendations with the Plant manager or QA manager, whether a third party or internal audit has been done, and if possible the auditor should give advice on how to solve any problems found.

At this stage we need to mention scoring. Many auditors or audit systems utilise a scoring system by which the findings of the audit are converted into a single score, expressed, for example, as percentage compliance or an approval grade (A, B, etc.). There are as many different scoring systems as there are audit methodologies, but each provides a means by which the results of the audit can be quickly and easily interpreted or compared by persons not involved in the audit. Scored audits also have the advantage in that, if the audits are carried out to the same standard, different audits can be compared quickly and easily, simply by using the score.

There are a number of points to remember when developing scoring systems for audits. The first is that, if not developed carefully, scoring systems can hide critical deficiencies. This can often happen if the scoring system allocates points for excellence or above standard. This immediately allows a company to overachieve in a section of the audit and to underachieve in another section, and when the results are averaged at the end they come out with a standard score. For this reason it is not advised to develop scoring systems which increase the score by overachieving; this should be rewarded in other ways.

There are several ways of ensuring that critical issues are accounted for in the overall score of the audit. The first is to have a weighting system, where the score for each question is multiplied by a weighting factor to give the final score for the question. The weight given to each question should reflect its contribution to product safety or quality. Thus, personnel wearing hair covers and overalls, whilst important, would not be weighted the same as having a calibrated cooking process and strict raw/cooked segregation in an area preparing cooked meats. Where trained auditors are used, it is possible to develop scoring systems where individual questions are not scored, but sections of the audit are scored. The score given to each section represents how well the company or department complies with the given standards, taking into account any critical areas covered in the section. The auditor is therefore looking at the overall picture, placing emphasis on critical issues when giving a score. This can be a very effective system but is obviously more subjective than the method mentioned above. It relies on having well-trained, experienced auditors, good standards and a well-developed audit questionnaire. This approach cannot be used with a check sheet. Where more than one auditor is used to carry out audits of this nature, it is also useful to set up a referee system, either by exchanging reports for discussion between auditors or by having the audit reports refereed by an experienced auditor to ensure consistency between auditors.

Follow-up
The food industry is ever changing. At the external level, new legislation and standards are introduced, new hazards, microbiological or chemical, are discovered which affect the way we work and the risks to our customers, and new process technologies become available. Within a business, new procedures are written, to take account of internal and external pressures, new processes are introduced and new products are manufactured. For this reason auditing cannot be 'one off'. For both internal and external auditing, regular audits are required in order to ensure that the systems and procedures keep pace with the external pressures on the business, and that internally, new procedures are implemented and effective.

Where an audit is part of an audit programme, follow-up is a vital part of ensuring that any actions resulting from a previous audit are being put into place.

Frequency
The frequency at which audits take place is dependent on the nature of the operation being audited. Major suppliers or suppliers of high risk ingredients (i.e. those which may carry pathogens or chemical contaminants) or finished packed product for direct sale will need to be audited more frequently than suppliers of minor ingredients.

7.3.2 Auditing HACCP systems

The principles described above are applicable for all types of audit. In the same way, auditing HACCP systems is no different from auditing other quality assurance systems such as ISO 9000. However, there are a number of points which should be considered. ISO 9000 as a system concentrates on the contractual relationship between supplier and customer, and the conformity to customer specifications.[1] The systems and procedures developed under the ISO 9000 system are therefore derived internally and specify the quality system for the whole company. HACCP differs from ISO 9000 in that it defines the hazards and controls related to a specific product or process, and a plant will have several different HACCP plans in place, one for each line/product, covering the total manufacturing operation. When auditing HACCP systems, therefore, the scope of the audit is likely to be very different from an ISO 9000-type audit.

Before auditing a HACCP system, it is important that the objective of the audit is very clear. HACCP audits make no check on the technical accuracy of the HACCP plan. This activity is part of the validation process which is discussed elsewhere in this book. A HACCP system audit is used to establish whether or not the controls, monitoring procedures and corrective actions defined in the HACCP plan are being applied correctly, and whether or not they are effective. It is a common misconception that HACCP audits will indicate whether a HACCP system is 'safe' and covers all applicable hazards. This is definitely not the case.

A HACCP systems audit would generally cover the following elements:

1. Have the HACCP studies been carried out according to the seven principles described by the Codex Alimentarius,[7] or an equivalent system?
2. Has a team approach been used to generate the HACCP plan, and what technical expertise has been available to the team?
3. Does that HACCP plan cover all the expected CCPs, together with targets, limits, monitoring systems and corrective actions? (This would normally be a part of validation, and would not be covered in an internal audit.)
4. Is there evidence that the HACCP plan has been validated?
5. Has the HACCP plan been discussed with operators, and do operators have access to work procedures based on HACCP? Have they been sufficiently trained and do they have sufficient tools and authority to carry out their responsibilities?
6. Are monitoring procedures being carried out and recorded on the factory floor? Is there any indication that the control procedures are not effective?
7. Are there clear priorities for action in the event of a process deviation?
8. Has the process changed since the study was carried out?
9. What verification data is available to demonstrate the effectiveness of the HACCP plan?
10. When was the HACCP plan last reviewed?

The above is not an exhaustive list but covers the main elements normally associated with a HACCP audit.

Internal auditing of a HACCP plan

In general there is very little difference in auditing a HACCP system in your own plant and in that of a third party. Both audits will require that the auditor assess the elements described above. However, in an audit 'in house' elements 1–4 above will be assessed initially and then left out of the regular audit system which would focus on elements 5–9. The key to auditing HACCP is not to spend a great deal of time examining the HACCP plan to check its accuracy – this will have been done when the plan was validated – but to focus on the operational side of HACCP. What we mean here is that the HACCP plan will define a number of controls and monitoring systems associated with each CCP. The aim of the audit is to check that working procedures are available which adequately cover the requirements at the CCP, that the operators have, and understand, these procedures and that any required data collected is being recorded and action taken if the process or material is outside the critical limits.

An important part of the HACCP audit is not only to check that the HACCP plan is implemented and the procedures are in place, but also to check that there have been no changes on the line, to working procedures (e.g. times, temperatures or hygiene) or to product formulation (e.g. preservation system or packaging) which may affect the effectiveness of the HACCP system. Although this is normally associated with the formal review of the HACCP system, it is normally not sensible to leave this type of check for the yearly review but to keep on top of the changes in this more frequent audit system.

External auditing of a HACCP plan

Auditing a third party HACCP plan follows the same principles as defined above. However, although it is not normally necessary to check the content and accuracy of your own HACCP plan, the auditor will need to make a judgement on the content and accuracy of the third party plan, to check its suitability for ensuring the safety of the supplied product. It is very difficult to assess another team's HACCP study, especially if you are not familiar with the product or the processes used by the third party. The way to tackle this problem is to identify the minimum CCPs that you would expect to find for the type of process being audited. This information can often be found in industry guides, or in generic HACCP plans which are produced for different sectors of the food industry. A note of caution here is that by their very nature these guides are generic and can be superficial. However, they should be of use in identifying the minimum number and location of CCPs which you should be able to find in the HACCP plan of the third party. If these minimum CCPs are not present this immediately warns the auditor that this HACCP plan is not likely to be effective at controlling the hazards in the process. As an aside, the Internet is a source of significant information with regard to HACCP, and many food companies post their HACCP plans on the Internet. These can be a useful source of information, but again they must be used carefully as these are individual company plans and have not undergone a peer group review, unlike the industry guides available. The USDA/FSIS provide a number of generic HACCP studies at http://www.inppaz.org.ar/MENUPAL/Bvirtual/FOS/haccp/usda/haccpmod.htm.

If the HACCP plan is acceptable, the auditor will then proceed to determine if the plan has been implemented in the factory and is working as intended.

7.4 What the auditor should look for

In any audit, time plays a crucial factor. The auditor never has sufficient time to cover all the elements they would like to, and good time management is critical to the success of an audit. As an auditor it is therefore important to remember that you will never be able to check everything, and should not try to do so. As a general guide the auditor should carry out following procedures.

- Start with a brief tour of the factory, starting with the raw materials and finishing where the finished goods leave. This tour is not a fact-finding exercise but is intended to give the auditor a general feel of the operation being audited. It will also provide an insight into the management attitude of the company with regard to quality and safety. A clean, tidy, well-organised factory with hand washing, clean operators with suitable protective clothing, notice-boards and signs instructing operators in good practice is always a good indication that the management are committed to quality and safety. On the other hand, an untidy, dirty and haphazardly organised factory gives a clear indication of a general disregard of the management for quality and safety. First impressions are significant, and although it is important that the auditor does not jump to too many conclusions from the initial visit, an experienced auditor will normally be able to tell what the outcome of the audit will be from this visit.
- The auditor should now check whether or not the required systems and procedures are in place to cover the required elements of the HACCP system, and whether they contain the necessary depth of information. The use of a well-designed check sheet or questionnaire is a vital aid to ensuring that all the relevant systems are covered during the audit. Remember that the auditor here is assessing against standards and not making a personal judgement.
- The existence of a well-written procedure is not an indication that the system is implemented in the company. It is the role of the auditor to check that what is written on paper is actually working and is effective. Although the auditor should check whether or not all the required procedures exist, they will not be able to verify that all procedures are in place and working. Therefore he or she should select a number of key elements to check. Selection of the elements to check should not be a random process and the auditor should always check a number of the CCPs defined in the HACCP plan, to assess whether what is described in the HACCP plan is in fact happening on the factory floor. This therefore involves checking that the work instructions for operators cover the work practices and any control measures and that the targets and limits are clearly specified to enable the operator to judge whether the CCP is in control. Monitoring procedures should be available on the line or in a laboratory and records should be meaningful, available and up to date.

- The auditor should also check a selection of other procedures so as to ensure the quality of implementation of the HACCP plan and the background of GMP. It is often useful, during the initial factory visit, to note any activities that do not appear to be in line with a given standard. If a procedure exists which covers the activity observed and which is not in accordance with the standard, clearly there is a problem with implementation.
- It is extremely important to talk to people, especially the operators on the production floor. It is possible to find out more about the current state of implementation of the company's quality system by talking to the operators than in any other way. (Do they know what a CCP is? Have they been told about HACCP?) Ask to see work procedures and line check sheets used for monitoring CCPs and other quality parameters. If the operator does not have the relevant procedure readily available, it is more than likely that the procedure is not being followed.

7.5 Future trends

Food quality and safety is continuously evolving and the foods industry needs to keep abreast of these changes to remain competitive and meet customer requirements. HACCP is here for the immediate future, and future trends in HACCP are discussed earlier in this book. However, one point to note is that many major customers now see HACCP as a key requirement from their suppliers, whereas in the past ISO 9000 was seen as the key requirement. As such, HACCP certification may become a more important feature of the HACCP system. Already many third party accreditation companies are offering HACCP certification services, either as a stand alone, or combined with existing ISO 9000 certification. International standards for assessing and certifying HACCP are being developed[16] with the aim of standardising the certification process. Currently, HACCP certification looks at the approach taken and the standards used for developing the HACCP plan and subsequent implementation of the plan. Technical accuracy of the HACCP plan will not usually be assessed and this may become a weakness of the certification process.

ISO 9000 was the dominant quality system in the early 1990s and is currently under revision (the so-called ISO 9000:2000 standards). This standard will retain the original 9001–9004 standards, but has changed the structure of the elements making up the standards. The ISO 9000:2000 standard has five elements, each with a number of sub-components:

- Quality Management System Requirements (one sub-component)
- Management Responsibility (six sub-components)
- Resource Management (three sub-components)
- Management of Processes (seven sub-components)
- Measurement, Analysis and Improvement (two sub-components).

Many of the sub-components are further subdivided. The 20 elements of the existing ISO 9000 system are covered within the five elements in the new system. However, the new system places more emphasis on validation and will hopefully address the issues associated with the current standard, whereby it is not inconceivable for a company to miss key activities within their internal system but be able to gain accreditation by demonstrating compliance with incorrect or incomplete standards defined internally.

For many companies it is difficult to find the resources, and the necessary skills, within their company for auditing third parties. In addition, many companies are faced with an increasing number of customer audits, which takes valuable resources from the day-to-day activities of the company. Third parties are picking up on these facts and offering third party auditing services, and even accredited auditing services. One such which is operational in the UK is the European Food Safety Inspection Service (EFSIS).[17] The system audits a plant against 35 set criteria in quality, safety and hygiene, and if the audit is acceptable will grant accreditation. Accreditation is a continual process and the frequency of re-accreditation will be determined depending on the type of process and the previous audit score. The rationale behind the EFSIS scheme is that it will reduce the number of third party or customer audits by providing third party auditors who will assess suppliers, and it will allow companies to show they have reached a set of fixed standards defined by EFSIS. Third party auditing and accreditation schemes are becoming seen as a good means of reducing the resource requirements in a company with regard to auditing, and offer independent assessment of a company's safety and quality system. Such systems are dependent upon the skills and professionalism of the auditors who carry out the assessments, but are likely to become more important in the food industry.

Current quality systems, and many of the associated auditing systems, focus on whether or not a system exists, and check that the system is actively implemented within the company. Very few systems require that the subsequent results of the implemented system are evaluated. Within Europe, the European Foundation for Quality Management (EFQM) has developed a model for quality excellence.[18,19] In common with ISO 9000 and HACCP, the EFQM system provides a framework for achieving excellence. This framework is built up of the following nine elements:

1. Leadership
2. People
3. Policy and strategy
4. Partnerships and resources
5. Processes
6. People results
7. Customer results
8. Society results
9. Key performance results.

However, unlike other systems, the EFQM divides these elements into 'enablers' and 'results', elements 1–5 being defined as enablers and elements 6–9 as results. Enablers are those criteria which define what the organisation does, and would be focused on internal policy, systems and procedures making up a quality system. The results are intended to cover what the organisation achieves, the premise being that there cannot be results without enablers. The EFQM website (http://www.efqm.org) describes the system in detail. The EFQM system is not the only system which focuses on results but the future lies with such systems, which look outside the organisation to ensure that what is defined internally has the desired results both internally and externally.

7.6 References

1. HARRIGAN W F, The ISO 9000 series and its implications for HACCP, *Food Control*, 1993 **4**(2) 105–11.
2. MAYES T, The application of management systems to food safety and quality, *Trends in Food Science & Technology*, 1993 **4**(7) 216–19.
3. ISO 9000, *Quality Management and Quality Assurance Standards – Guidelines for Selection and Use*, 1987.
4. ISO 9001, *Quality Systems – Model for Quality Assurance in Design/ Development, Production, Installation and Servicing*, 1987.
5. ISO 9002, *Quality Systems – Model for Quality Assurance in Production and Installation*, 1987.
6. ISO 9003, *Quality Systems – Model for Quality Assurance in Final Inspection and Test*, 1987.
7. ISO 9004, *Quality Management and Quality System Elements – Guidelines*, 1987.
8. Codex Alimentarius, *Hazards Analysis and Critical Control Point System and Guidelines for its Application*, Alinorm 97/13A, Codex Alimentarius Commission, Rome, 1997.
9. OAKLAND J S, *Total Quality Management. The route to improving performance*, 2nd edn, Oxford, Butterworth Heinemann, 1995.
10. British Standard Quality Systems (BS 5750), British Standards Institution.
11. JOUVE J L, STRINGER M F and BAIRD-PARKER A C, Food safety management tools, *Food Science and Technology Today*, 1999 **13**(2) 82–91.
12. *Recommended International Code of Practice – General Principles of Food Hygiene*, CAC/RCP–1 (1969), rev.3 (1997).
13. Institute of Food Science and Technology, *Food and Drink Good Manufacturing Practice Guidelines* (a guide to its responsible management), 4th edn, 1998.
14. CHESWORTH N, Implementing a factory auditing programme. *Int. Food Hygiene*, 1993 **4**(4) 11–13.
15. Codex Alimentarius, *Food Hygiene, Basic Texts*. Joint FAO/WHO Food Standards Programme, FAO Rome, ISBN 92-5-104021-4, 1997.

16. Criteria for testing an operational HACCP system, Central Board of HACCP Experts, The Hague, PO Box 93093, 2509 AB, The Hague, 1996.
17. RICHARDSON D, Audit after audit, is there an alternative? *Food Manufacture*, 1998 **70**(4) 20.
18. ROGERS V, EFQM, a model for management excellence, *Food Manufacture*, 1998 **73**(12) 20.
19. The EFQM Excellence Model, http://www.efqm.org

8

Laboratories and analytical methods

Quality control

R. Wood, Food Standards Agency, London

8.1 Introduction

It is now universally recognised as being essential that a laboratory produces and reports data that are fit-for-purpose. For a laboratory to produce consistently reliable data it must implement an appropriate programme of quality assurance measures; such measures are now required by virtue of legislation for food control work or, in the case of the UK Food Standards Agency (FSA), in their requirements for contractors undertaking surveillance work. Thus customers now demand of providers of analytical data that their data meet established quality requirements. These are further described below. The significance of the measures identified are then described and some indications are given as to the future of analytical methodology within the food laboratory. These are then discussed. The need to meet established quality requirements and often implement new experimental protocols will demand that laboratories prepare to be audited. The increasingly detailed nature of relevant guidance documents such as ISO/IEC/17025 will result in more rigorous audit procedures.

Methods of analysis have been prescribed by legislation for a number of foodstuffs since the UK acceded to the European Community in 1972. However, the Community now recognises that the quality of results from a laboratory is equally as important as the method used to obtain the results. This is best illustrated by consideration of the Council Directive on the Official Control of Foodstuffs (OCF) which was adopted by the Community in June, 1989.[1] This, and the similar Codex Alimentarius Commission requirements, are described below. As a result of this general recognition there is a general move away from the need to prescribe all analytical methodology in detail towards the prescription of the general quality systems within which the laboratory must operate. This allows greater flexibility to the laboratory without detracting from the quality of results that it will produce. This shift to setting up appropriate quality systems has raised the profile of auditing as a means of assessing such systems. This chapter is designed to describe the key features of such quality systems that both laboratories and auditors need to take account of.

8.2 Legislative requirements

8.2.1 The European Union

For analytical laboratories in the food sector there are legislative requirements regarding analytical data which have been adopted by the European Union. In particular, methods of analysis have been prescribed by legislation for a number of foodstuffs since the UK acceded to the European Community in 1972. However, the Union now recognises that the competency of a laboratory (i.e. how well it can use a method) is equally as important as the 'quality' of the method used to obtain results. The Council Directive on the Official Control of Foodstuffs which was adopted by the Community in 1989[1] looks forward to the establishment of laboratory quality standards, by stating that 'In order to ensure that the application of this Directive is uniform throughout the Member States, the Commission shall, within one year of its adoption, make a report to the European Parliament and to the Council on the possibility of establishing Community quality standards for all laboratories involved in inspection and sampling under this Directive' (Article 13).

Following that, in September 1990 the Commission produced a Report which recommended establishing Community quality standards for all laboratories involved in inspections and sampling under the OCF Directive. Proposals on this have now been adopted by the Community in the Directive on Additional Measures Concerning the Food Control of Foodstuffs (AMFC).[2] In Article 3 of the AMFC Directive it states:

1. Member States shall take all measures necessary to ensure that the laboratories referred to in Article 7 of Directive 89/397/EEC[1] comply with the general criteria for the operation of testing laboratories laid down in European standard EN 45001[3] supplemented by Standard Operating Procedures and the random audit of their compliance by quality assurance personnel, in accordance with the OECD principles Nos. 2 and 7 of good laboratory practice as set out in Section II of Annex 2 of the Decision of the Council of the OECD of 12 Mar 1981 concerning the mutual acceptance of data in the assessment of chemicals.[4]
2. In assessing the laboratories referred to in Article 7 of Directive 89/397/EEC Member States shall:
   (a) apply the criteria laid down in European standard EN 45002;[5] and
   (b) require the use of proficiency testing schemes as far as appropriate.

Laboratories meeting the assessment criteria shall be presumed to fulfil the criteria referred to in paragraph 1. Laboratories which do not meet the assessment criteria shall not be considered as laboratories referred to in Article 7 of the said Directive.

3. Member States shall designate bodies responsible for the assessment of laboratories as referred to in Article 7 of Directive 89/397/EEC. These bodies shall comply with the general criteria for laboratory accreditation bodies laid down in European Standard EN 45003.[6]
4. The accreditation and assessment of testing laboratories referred to in this article may relate to individual tests or groups of tests. Any appropriate deviation in the way in which the standards referred to in paragraphs 1, 2 and 3 are applied shall be adopted in accordance with the procedure laid down in Article 8.

In Article 4, it states:

Member States shall ensure that the validation of methods of analysis used within the context of official control of foodstuffs by the laboratories referred to in Article 7 of Directive 89/397/EEC comply whenever possible with the provisions of paragraphs 1 and 2 of the Annex to Council Directive 85/591/EEC of 23 December 1985 concerning the introduction of Community methods of sampling and analysis for the monitoring of foodstuffs intended for human consumption.[7]

As a result of the adoption of the above directives legislation is now in place to ensure that there is confidence not only in national laboratories but also those of the other Member States. As one of the objectives of the EU is to promote the concept of mutual recognition, this is being achieved in the laboratory area by the adoption of the AMFC directive. The effect of the AMFC Directive is that organisations must consider the following aspects within the laboratory: its organisation, how well it actually carries out analyses, and the methods of analysis used in the laboratory. All these aspects are inter-related, but in simple terms may be thought of as:

- becoming accredited to an internationally recognised standard; such accreditation is aided by the use of internal quality control procedures
- participating in proficiency schemes, and
- using validated methods.

In addition it is important that there is dialogue and co-operation by the laboratory with its customers. This is also required by virtue of the EN 45001 Standard at paragraph 6, and will be emphasised even more in future with the adoption of ISO/IEC Guide 17025.[8]

The AMFC Directive requires that food control laboratories should be accredited to the EN 45000 series of standards as supplemented by some of the OECD GLP principles. In the UK, government departments have nominated the United Kingdom Accreditation Service (UKAS) to carry out the accreditation of official food control laboratories for all the aspects prescribed in the Directive. However, as the accreditation agency will also be required to comply with the EN

45003 Standard and to carry out assessments in accordance with the EN 45002 Standard, all accreditation agencies that are members of the European Co-operation for Accreditation of Laboratories (EA) may be asked to carry out the accreditation of a food control laboratory within the UK. Similar procedures will be followed in the other Member States, all having or developing equivalent organisations to UKAS. Details of the UK requirements for food control laboratories are described later in this chapter.

8.2.2 Codex Alimentarius Commission: guidelines for the assessment of the competence of testing laboratories involved in the import and export control of food

The decisions of the Codex Alimentarius Commission are becoming of increasing importance because of the acceptance of Codex Standards in the World Trade Organisation agreements. They may be regarded as being semi-legal in status. Thus, on a world-wide level, the establishment of the World Trade Organisation (WTO) and the formal acceptance of the Agreements on the Application of Sanitary and Phytosanitary Measures (SPS Agreement) and Technical Barriers to Trade (TBT Agreement) have dramatically increased the status of Codex as a body. As a result, Codex Standards are now seen as *de facto* international standards and are increasingly being adopted by reference into the food law of both developed and developing countries.

Because of the status of the CAC described above, the work that it has carried out in the area of laboratory quality assurance must be carefully considered. One of the CAC Committees, the Codex Committee on Methods of Analysis and Sampling (CCMAS), has developed criteria for assessing the competence of testing laboratories involved in the official import and export control of foods. These were recommended by the Committee at its 21st Session in March 1997[9] and adopted by the Codex Alimentarius Commission at its 22nd Session in June 1997;[10] they mirror the EU recommendations for laboratory quality standards and methods of analysis. The guidelines provide a framework for the implementation of quality assurance measures to ensure the competence of testing laboratories involved in the import and export control of foods. They are intended to assist countries in their fair trade in foodstuffs and to protect consumers.

The criteria for laboratories involved in the import and export control of foods, now adopted by the Codex Alimentarius Commission, are:

- to comply with the general criteria for testing laboratories laid down in ISO/IEC Guide 25: 1990 'General requirements for the competence of calibration and testing laboratories'[8] (i.e. effectively accreditation)
- to participate in appropriate proficiency testing schemes for food analysis which conform to the requirements laid down in 'The International Harmonised Protocol for the Proficiency Testing of (Chemical) Analytical Laboratories'[11] (already adopted for Codex purposes by the CAC at its 21st Session in July 1995)

- to use, whenever available, methods of analysis that have been validated according to the principles laid down by the CAC
- to use internal quality control procedures, such as those described in the 'Harmonised Guidelines for Internal Quality Control in Analytical Chemistry Laboratories'.[12]

In addition, the bodies assessing the laboratories should comply with the general criteria for laboratory accreditation, such as those laid down in the ISO/IEC Guide 58:1993: 'Calibration and testing laboratory accreditation systems – General requirements for operation and recognition'.[13]

Thus, as for the European Union, the requirements are based on accreditation, proficiency testing, the use of validated methods of analysis and, in addition, the formal requirement to use internal quality control procedures which comply with the Harmonised Guidelines. Although the EU and Codex Alimentarius Commission refer to different sets of accreditation standards, the ISO/IEC Guide 25: 1990 and EN 45000 series of standards are similar in intent. It is only through these measures that international trade will be facilitated and the requirements to allow mutual recognition to be fulfilled will be achieved. The EU or other relevant trading partners will scrutinise competent authorities which will involve an audit of their approved laboratories. It is essential that these laboratories operate rigorous control systems which meet approved standards. The ISO/IEC guide 17025 provides instructions to the laboratory on building such a system. Section 4.13 highlights the importance of internal audit. This guide contains specific guidance on technical competence requirements not covered by ISO 9001/2. Further guidance on auditing is available from ISO in document series 10011-1 to 1011-3 which gives guidance to auditing and managing quality programmes. Further technical details are available in the bibliography of ISO/IEC 17025 for laboratories strengthening their internal control systems – particularly the aspects of accuracy, precision, proficiency and uncertainty.

8.3 FSA surveillance requirements

The Food Standards Agency undertakes food chemical surveillance exercises. It has developed information for potential contractors on the analytical quality assurance requirements for food chemical surveillance exercises. These requirements are described below but reproduced as an appendix to this chapter; they emphasise the need for a laboratory to produce and report data of appropriate quality. The requirements are divided into three parts dealing with:

- Part A: quality assurance requirements for surveillance projects provided by potential contractors at the time that tender documents are completed and when commissioning a survey. Here information is sought on:
  - the formal quality system in the laboratory if third-party assessed (i.e. if UKAS accredited or GLP compliant)

- the quality system if not accredited
- proficiency testing
- internal quality control
- method validation.

- Part B: information to be defined by the FSA customer once the contract has been awarded to a contractor, e.g. the sample storage conditions to be used, the methods to be used, the IQC procedures to be used, the required measurement limits (e.g. limit of detection (LOD), limit of determination/quantification (LOQ), and the reporting limits)
- Part C: information to be provided by the contractor on an on-going basis once contract is awarded – to be agreed with the customer to ensure that the contractor remains in 'analytical control'.

8.4 Laboratory accreditation and quality control

Although the legislative requirements apply only to food-control laboratories, the effect of their adoption is that other food laboratories will be advised to achieve the same standard in order for their results to be recognised as equivalent and accepted for 'due diligence' purposes. In addition, the Codex requirements affect all organisations involved in international trade and thus provide an important 'quality umbrella'.

As shown above, these include a laboratory to be third-party assessed to international accreditation standards, to demonstrate that it is in statistical control by using appropriate internal quality control procedures, to participate in proficiency testing schemes which provide an objective means of assessing and documenting the reliability of the data it is producing and to use methods of analysis that are 'fit-for-purpose'. These requirements are summarised below and then described in greater detail later in this chapter.

8.4.1 Accreditation: preparing a laboratory for audit

The AMFC Directive requires that food-control laboratories should be accredited to the EN 45000 series of standards as supplemented by some of the OECD GLP principles. In the UK, government departments will nominate the United Kingdom Accreditation Service (UKAS) to carry out the accreditation of official food-control laboratories for all the aspects prescribed in the Directive. However, as the accreditation agency will also be required to comply to EN 45003 Standard and to carry out assessments in accordance with the EN 45002 Standard, any other accreditation agencies that are members of the European Co-operation for Accreditation of Laboratories (EA) may also be nominated to carry out the accreditation. Similar procedures will be followed in the other Member States, all having or developing equivalent organisations to UKAS.

It has been the normal practice for UKAS to accredit the scope of laboratories on a method-by-method basis. In the case of official food-control laboratories undertaking non-routine or investigative chemical analysis it is accepted that it is not practical to use an accredited fully documented method in the conventional sense, which specifies each sample type and analyte. In these cases a laboratory must have a protocol defining the approach to be adopted which includes the requirements for validation and quality control. Full details of procedures used, including instrumental parameters, must be recorded at the time of each analysis in order to enable the procedure to be repeated in the same manner at a later date. It is therefore recommended that for official food-control laboratories undertaking analysis, appropriate methods are accredited on a generic basis with such generic accreditation being underpinned where necessary by specific method accreditation.

Food-control laboratories seeking to be accredited for the purposes of the Directive should include, as a minimum, the following techniques in generic protocols: HPLC, GC, atomic absorption and/or ICP (and microscopy). A further protocol on sample preparation procedures (including digestion and solvent dissolution procedures) should also be developed. Other protocols for generic methods which are acceptable to UKAS may also be developed. Proximate analyses should be addressed as a series of specific methods including moisture, fat, protein and ash determinations.

Where specific Regulations are in force then the methods associated with the Regulations shall be accredited if the control laboratory wishes to offer enforcement of the Regulations to customers. Examples of these are methods of analysis for aflatoxins and methods of analysis for specific and overall migration for food contact materials.

By using the combination of specific method accreditation and generic accreditation it will be possible for laboratories to be accredited for all the analyses of which they are capable and competent to undertake. Method performance validation data demonstrating that the method was fit-for-purpose shall be demonstrated before the test result is released and method performance shall be monitored by on-going quality-control techniques where applicable. It will be necessary for laboratories to be able to demonstrate quality-control procedures to ensure compliance with the EN 45001 Standard,[3] an example of which would be compliance with the ISO/AOAC/IUPAC Guidelines on Internal Quality Control in Analytical Chemistry Laboratories.[12]

8.4.2 Internal quality control (IQC)

IQC is one of a number of concerted measures that analytical chemists can take to ensure that the data produced in the laboratory are of known quality and uncertainty. In practice this is determined by comparing the results achieved in the laboratory at a given time with a standard. IQC therefore comprises the routine practical procedures that enable the analyst to accept a result or group of results or reject the results and repeat the analysis. IQC is undertaken by the inclusion of particular reference materials, 'control materials', into the analytical sequence and by duplicate analysis.

ISO, IUPAC and AOAC INTERNATIONAL have co-operated to produce agreed protocols on the 'Design, Conduct and Interpretation of Collaborative Studies'[14] and on the 'Proficiency Testing of [Chemical] Analytical Laboratories'.[11] The Working Group that produced these protocols has prepared a further protocol on the internal quality control of data produced in analytical laboratories. The document was finalised in 1994 and published in 1995 as the 'Harmonised Guidelines For Internal Quality Control In Analytical Chemistry Laboratories'.[12] The use of the procedures outlined in the Protocol should aid compliance with the accreditation requirements specified above. The successful demonstration of these control procedures will result in the laboratory 'passing' mandatory external audits from an approved authority or from market-led assessments by designated third party auditors.

Basic concepts

The protocol sets out guidelines for the implementation of internal quality control (IQC) in analytical laboratories. IQC is one of a number of concerted measures that analytical chemists can take to ensure that the data produced in the laboratory are fit for their intended purpose. In practice, fitness for purpose is determined by a comparison of the accuracy achieved in a laboratory at a given time with a required level of accuracy. Internal quality control therefore comprises the routine practical procedures that enable the analytical chemist to accept a result or group of results as fit-for-purpose, or reject the results and repeat the analysis. As such, IQC is an important determinant of the quality of analytical data, and is recognised as such by accreditation agencies.

Internal quality control is undertaken by the inclusion of particular reference materials, called 'control materials', into the analytical sequence and by duplicate analysis. The control materials should, wherever possible, be representative of the test materials under consideration in respect of matrix composition, the state of physical preparation and the concentration range of the analyte. As the control materials are treated in exactly the same way as the test materials, they are regarded as surrogates that can be used to characterise the performance of the analytical system, both at a specific time and over longer intervals. Internal quality control is a final check of the correct execution of all of the procedures (including calibration) that are prescribed in the analytical protocol and all of the other quality assurance measures that underlie good analytical practice. IQC is therefore necessarily retrospective. It is also required to be as far as possible independent of the analytical protocol, especially the calibration, that it is designed to test.

Ideally both the control materials and those used to create the calibration should be traceable to appropriate certified reference materials or a recognised empirical reference method. When this is not possible, control materials should be traceable at least to a material of guaranteed purity or other well characterised material. However, the two paths of traceability must not become coincident at too late a stage in the analytical process. For instance, if control materials and calibration standards were prepared from a single stock solution of analyte, IQC would not detect any inaccuracy stemming from the incorrect preparation of the stock solution.

In a typical analytical situation several, or perhaps many, similar test materials will be analysed together, and control materials will be included in the group. Often determinations will be duplicated by the analysis of separate test portions of the same material. Such a group of materials is referred to as an analytical 'run'. (The words 'set', 'series' and 'batch' have also been used as synonyms for 'run'.) Runs are regarded as being analysed under effectively constant conditions. The batches of reagents, the instrument settings, the analyst, and the laboratory environment will, under ideal conditions, remain unchanged during analysis of a run. Systematic errors should therefore remain constant during a run, as should the values of the parameters that describe random errors. As the monitoring of these errors is of concern, the run is the basic operational unit of IQC.

A run is therefore regarded as being carried out under repeatability conditions, i.e. the random measurement errors are of a magnitude that would be encountered in a 'short' period of time. In practice the analysis of a run may occupy sufficient time for small systematic changes to occur. For example, reagents may degrade, instruments may drift, minor adjustments to instrumental settings may be called for, or the laboratory temperature may rise. However, these systematic effects are, for the purposes of IQC, subsumed into the repeatability variations. Sorting the materials making up a run into a randomised order converts the effects of drift into random errors.

Scope of the guidelines

The guidelines are a harmonisation of IQC procedures that have evolved in various fields of analysis, notably clinical biochemistry, geochemistry, environmental studies, occupational hygiene and food analysis. There is much common ground in the procedures from these various fields. However, analytical chemistry comprises an even wider range of activities, and the basic principles of IQC should be able to encompass all of these. The guidelines will be applicable in the great majority of instances although there are a number of IQC practices that are restricted to individual sectors of the analytical community and so not included in the guidelines.

In order to achieve harmonisation and provide basic guidance on IQC, some types of analytical activity have been excluded from the guidelines. Issues specifically excluded are as follows:

- *Quality control of sampling.* While it is recognised that the quality of the analytical result can be no better than that of the sample, quality control of sampling is a separate subject and in many areas not yet fully developed. Moreover, in many instances analytical laboratories have no control over sampling practice and quality.
- *In-line analysis and continuous monitoring.* In this style of analysis there is no possibility of repeating the measurement, so the concept of IQC as used in the guidelines is inapplicable.

- *Multivariate IQC.* Multivariate methods in IQC are still the subject of research and cannot be regarded as sufficiently established for inclusion in the guidelines. The current document regards multianalyte data as requiring a series of univariate IQC tests. Caution is necessary in the interpretation of this type of data to avoid inappropriately frequent rejection of data.
- *Statutory and contractual requirements.*
- *Quality assurance measures* such as pre-analytical checks on instrumental stability, wavelength calibration, balance calibration, tests on resolution of chromatography columns, and problem diagnostics are not included. For present purposes they are regarded as part of the analytical protocol, and IQC tests their effectiveness together with the other aspects of the methodology.

Recommendations

The following recommendations represent integrated approaches to IQC that are suitable for many types of analysis and applications areas. Managers of laboratory quality systems will have to adapt the recommendations to the demands of their own particular requirements. Such adoption could be implemented, for example, by adjusting the number of duplicates and control material inserted into a run, or by the inclusion of any additional measures favoured in the particular application area. The procedure finally chosen and its accompanying decision rules must be codified in an IQC protocol that is separate from the analytical system protocol. The practical approach to quality control is determined by the frequency with which the measurement is carried out and the size and nature of each run. The following recommendations are therefore made. (The use of control charts and decision rules are covered in Appendix 1 to the guidelines.)

In all of the following the order in the run in which the various materials are analysed should be randomised if possible. A failure to randomise may result in an underestimation of various components of error.

Short (e.g. $n < 20$) frequent runs of similar materials

Here the concentration range of the analyte in the run is relatively small, so a common value of standard deviation can be assumed. Insert a control material at least once per run. Plot either the individual values obtained, or the mean value, on an appropriate control chart. Analyse in duplicate at least half of the test materials, selected at random. Insert at least one blank determination

Longer (e.g. $n > 20$) frequent runs of similar materials

Again a common level of standard deviation is assumed. Insert the control material at an approximate frequency of one per ten test materials. If the run size is likely to vary from run to run it is easier to standardise on a fixed number of insertions per run and plot the mean value on a control chart of means. Otherwise plot individual values. Analyse in duplicate a minimum of five test materials selected at random. Insert one blank determination per ten test materials.

Frequent runs containing similar materials but with a wide range of analyte concentration
Here we cannot assume that a single value of standard deviation is applicable. Insert control materials in total numbers approximately as recommended above. However, there should be at least two levels of analyte represented, one close to the median level of typical test materials, and the other approximately at the upper or lower decile as appropriate. Enter values for the two control materials on separate control charts. Duplicate a minimum of five test materials, and insert one procedural blank per ten test materials.

Ad hoc analysis
Here the concept of statistical control is not applicable. It is assumed, however, that the materials in the run are of a single type. Carry out duplicate analysis on all of the test materials. Carry out spiking or recovery tests or use a formulated control material, with an appropriate number of insertions (see above), and with different concentrations of analyte if appropriate. Carry out blank determinations. As no control limits are available, compare the bias and precision with fitness-for-purpose limits or other established criteria.

By following the above recommendations laboratories would introduce internal quality control measures which are an essential aspect of ensuring that data released from a laboratory are fit-for-purpose. If properly executed, quality control methods can monitor the various aspects of data quality on a run-by-run basis. In runs where performance falls outside acceptable limits, the data produced can be rejected and, after remedial action on the analytical system, the analysis can be repeated.

The guidelines stress, however, that internal quality control is not foolproof even when properly executed. Obviously it is subject to 'errors of both kinds', i.e. runs that are in control will occasionally be rejected and runs that are out of control occasionally accepted. Of more importance, IQC cannot usually identify sporadic gross errors or short-term disturbances in the analytical system that affect the results for individual test materials. Moreover, inferences based on IQC results are applicable only to test materials that fall within the scope of the analytical method validation. Despite these limitations, which professional experience and diligence can alleviate to a degree, internal quality control is the principal recourse available for ensuring that only data of appropriate quality are released from a laboratory. When properly executed it is very successful.

The guidelines also stress that the perfunctory execution of any quality system will not guarantee the production of data of adequate quality. The correct procedures for feedback, remedial action and staff motivation must also be documented and acted upon. In other words, there must be a genuine commitment to quality within a laboratory for an internal quality control programme to succeed, i.e. the IQC must be part of a complete quality management system.

8.5 Proficiency testing

Participation in proficiency testing schemes provides laboratories with an objective means of assessing and documenting the reliability of the data they are producing. Although there are several types of proficiency testing schemes they all share a common feature: test results obtained by one laboratory are compared with those obtained by one or more testing laboratories. The proficiency testing schemes must provide a transparent interpretation and assessment of results. Laboratories wishing to demonstrate their proficiency should seek and participate in proficiency testing schemes relevant to their area of work.

The need for laboratories carrying out analytical determinations to demonstrate that they are doing so competently has become paramount. It may well be necessary for such laboratories not only to become accredited and to use fully validated methods but also to participate successfully in proficiency testing schemes. Thus, proficiency testing has assumed a far greater importance than previously.

8.5.1 What is proficiency testing?

A proficiency testing scheme is defined as a system for objectively checking laboratory results by an external agency. It includes comparison of a laboratory's results at intervals with those of other laboratories, the main object being the establishment of trueness. In addition, although various protocols for proficiency testing schemes have been produced the need now is for a harmonised protocol that will be universally accepted; the progress towards the preparation and adoption of an internationally recognised protocol is described below. Various terms have been used to describe schemes conforming to the protocol (e.g. external quality assessment, performance schemes, etc.), but the preferred term is 'proficiency testing'.

Proficiency testing schemes are based on the regular circulation of homogeneous samples by a co-ordinator, analysis of samples (normally by the laboratory's method of choice) and an assessment of the results. However, although many organisations carry out such schemes, there has been no international agreement on how this should be done – in contrast to the collaborative trial situation. In order to rectify this, the same international group that drew up collaborative trial protocols was invited to prepare one for proficiency schemes (the first meeting to do so was held in April 1989). Other organisations, such as CEN, are also expected to address the problem.

8.5.2 Why proficiency testing is important

Participation in proficiency testing schemes provides laboratories with a means of objectively assessing, and demonstrating, the reliability of the data they produce. Although there are several types of schemes, they all share a common feature of comparing test results obtained by one testing laboratory with those obtained by other testing laboratories. Schemes may be 'open' to any laboratory or participation may be invited. Schemes may set out to assess the competence of laboratories undertaking a very specific analysis (e.g. lead in blood) or more general analysis (e.g. food analysis). Although accreditation and proficiency testing are separate exercises, it is anticipated that accreditation assessments will increasingly use proficiency testing data.

8.5.3 Accreditation agencies

It is now recommended by ISO Guide 25,[8] the prime standard to which accreditation agencies operate, that such agencies require laboratories seeking accreditation to participate in an appropriate proficiency testing scheme before accreditation is gained. There is now an internationally recognised protocol to which proficiency testing schemes should comply; this is the IUPAC/AOAC/ISO Harmonised Protocol described below.

8.5.4 ISO/IUPAC/AOAC International Harmonised Protocol For Proficiency Testing of (Chemical) Analytical Laboratories

The International Standardising Organisations, AOAC, ISO and IUPAC, have co-operated to produce an agreed 'International Harmonised Protocol For Proficiency Testing of (Chemical) Analytical Laboratories'.[11] That protocol is recognised within the food sector of the European Community and also by the Codex Alimentarius Commission. The protocol makes the following recommendations about the organisation of proficiency testing, all of which are important in the food sector.

Framework

Samples must be distributed regularly to participants who are to return results within a given time. The results will be statistically analysed by the organiser and participants will be notified of their performance. Advice will be available to poor performers and participants will be kept fully informed of the scheme's progress. Participants will be identified by code only, to preserve confidentiality. The scheme's structure for any one analyte or round in a series should be:

- samples prepared
- samples distributed regularly
- participants analyse samples and report results
- results analysed and performance assessed
- participants notified of their performance
- advice available for poor performers, on request
- co-ordinator reviews performance of scheme
- next round commences.

Organisation

The running of the scheme will be the responsibility of a co-ordinating laboratory/organisation. Sample preparation will either be contracted out or undertaken in house. The co-ordinating laboratory must be of high reputation in the type of analysis being tested. Overall management of the scheme should be in the hands of a small steering committee (Advisory Panel) having representatives from the co-ordinating laboratory (who should be practising laboratory scientists), contract laboratories (if any), appropriate professional bodies and ordinary participants.

Samples

The samples to be distributed must be generally similar in matrix to the unknown samples that are routinely analysed (in respect of matrix composition and analyte concentration range). It is essential they are of acceptable homogeneity and stability. The bulk material prepared must be effectively homogeneous so that all laboratories will receive samples that do not differ significantly in analyte concentration. The co-ordinating laboratory should also show the bulk sample is sufficiently stable to ensure it will not undergo significant change throughout the duration of the proficiency test. Thus, prior to sample distribution, matrix and analyte stability must be determined by analysis after appropriate storage. Ideally, the quality checks on samples referred should be performed by a different laboratory from that which prepared the sample, although it is recognised that this would probably cause considerable difficulty to the co-ordinating laboratory. The number of samples to be distributed per round for each analyte should be no more than five.

Frequency of sample distribution

Sample distribution frequency in any one series should not be more than every two weeks and not less than every four months. A frequency greater than once every two weeks could lead to problems in turn-round of samples and results. If the period between distributions extends much beyond four months, there will be unacceptable delays in identifying analytical problems and the impact of the scheme on participants will be small. The frequency also relates to the field of application and amount of internal quality control that is required for that field. Thus, although the frequency range stated above should be adhered to, there may be circumstances where it is acceptable for a longer time scale between sample distribution, e.g. if sample throughput per annum is very low. Advice on this respect would be a function of the Advisory Panel.

Estimating the assigned value (the 'true' result)

There are a number of possible approaches to determining the nominally 'true' result for a sample but only three are normally considered. The result may be established from the amount of analyte added to the samples by the laboratory preparing the sample; alternatively, a 'reference' laboratory (or group of such expert laboratories) may be asked to measure the concentration of the analyte using definitive methods or thirdly, the results obtained by the participating laboratories (or a substantial sub-group of these) may be used as the basis for the nominal 'true' result. The organisers of the scheme should provide the participants with a clear statement giving the basis for the assignment of reference values which should take into account the views of the Advisory Panel.

Choice of analytical method
Participants can use the analytical method of their choice except when otherwise instructed to adopt a specified method. It is recommended that all methods should be properly validated before use. In situations where the analytical result is method-dependent the true value will be assessed using those results obtained using a defined procedure. If participants use a method that is not 'equivalent' to the defining method, then an automatic bias in result will occur when their performance is assessed.

Performance criteria
For each analyte in a round a criterion for the performance score may be set, against which the score obtained by a laboratory can be judged. A 'running score' could be calculated to give an assessment of performance spread over a longer period of time.

Reporting results
Reports issued to participants should include data on the results from all laboratories together with participants' own performance score. The original results should be presented to enable participants to check correct data entry. Reports should be made available before the next sample distribution. Although all results should be reported, it may not be possible to do this in very extensive schemes (e.g. 800 participants determining 15 analytes in a round). Participants should, therefore, receive at least a clear report with the results of all laboratories in histogram form.

Liaison with participants
Participants should be provided with a detailed information pack on joining the scheme. Communication with participants should be by newsletter or annual report together with a periodic open meeting; participants should be advised of changes in scheme design. Advice should be available to poor performers. Feedback from laboratories should be encouraged so participants contribute to the scheme's development. Participants should view it as their scheme rather than one imposed by a distant bureaucracy.

Collusion and falsification of results
Collusion might take place between laboratories so that independent data are not submitted. Proficiency testing schemes should be designed to ensure that there is as little collusion and falsification as possible. For example, alternative samples could be distributed within a round. Also instructions should make it clear that collusion is contrary to professional scientific conduct and serves only to nullify the benefits of proficiency testing.

8.5.5 Statistical procedure for the analysis of results

The first stage in producing a score from a result x (a single measurement of analyte concentration in a test material) is to obtain an estimate of the bias, thus:

$$\text{bias} = x - X$$

where X is the true concentration or amount of analyte.

The efficacy of any proficiency test depends on using a reliable value for X. Several methods are available for establishing a working estimate of $\hat{X}$ (i.e. the assigned value).

Formation of a z-score

Most proficiency testing schemes compare bias with a standard error. An obvious approach is to form the z-score given by:

$$z = (x - \hat{X})/\sigma$$

where σ is a standard deviation. σ could be either an estimate of the actual variation encountered in a particular round ($\tilde{s}$) estimated from the laboratories' results after outlier elimination or a target representing the maximum allowed variation consistent with valid data.

A fixed target value for σ is preferable and can be arrived at in several ways. It could be fixed arbitrarily, with a value based on a perception of how laboratories should perform. It could be an estimate of the precision required for a specific task of data interpretation. σ could be derived from a model of precision, such as the 'Horwitz Curve'.[15] However, while this model provides a general picture of reproducibility, substantial deviation from it may be experienced for particular methods.

Interpretation of z-scores

If $\hat{X}$ and σ are good estimates of the population mean and standard deviation then z will be approximately normally distributed with a mean of zero and unit standard deviation. An analytical result is described as 'well behaved' when it complies with this condition.

An absolute value of z ($|z|$) greater than three suggests poor performance in terms of accuracy. This judgement depends on the assumption of the normal distribution, which, outliers apart, seems to be justified in practice.

As z is standardised, it is comparable for all analytes and methods. Thus values of z can be combined to give a composite score for a laboratory in one round of a proficiency test. The z-scores can therefore be interpreted as follows:

$|z| < 2$ 'Satisfactory': will occur in 95% of cases produced by 'well behaved results'

$2 < |z| < 3$ 'Questionable': but will occur in $\approx$5% of cases produced by 'well behaved results'

$|z| > 3$ 'Unsatisfactory': will only occur in $\approx$0.1% of cases produced by 'well behaved results'

Combination of results within a round of the trial
There are several methods of combining the z-scores produced by a laboratory in one round of the proficiency test described in the Protocol. They are:

The sum of scores, $SZ = \Sigma z$

The sum of squared scores, $SSZ = \Sigma z^2$

The sum of absolute values of the scores, $SAZ = \Sigma\ |z|$

All should be used with caution, however. It is the individual z-scores that are the critical consideration when considering the proficiency of a laboratory.

Calculation of running scores
Similar considerations apply for running scores as apply to combination scores above.

8.6 Analytical methods

Analytical methods should be validated as fit-for-purpose before use by a laboratory. Laboratories should ensure that, as a minimum, the methods they use are fully documented, laboratory staff trained in their use and control mechanisms established to ensure that the procedures are under statistical control.

The development of methods of analysis for incorporation into International Standards or into foodstuff legislation was, until comparatively recently, not systematic. However, the EU and Codex have requirements regarding methods of analysis and these are outlined below. They are followed by other International Standardising Organisations (e.g. AOAC International (AOACI) and the European Committee for Standardization (CEN)).

8.6.1 Codex Alimentarius Commission

This was the first international organisation working at the government level in the food sector that laid down principles for the establishment of its methods. That it was necessary for such guidelines and principles to be laid down reflects the confused and unsatisfactory situation in the development of legislative methods of analysis that existed until the early 1980s in the food sector.

The 'Principles for the Establishment of Codex Methods of Analysis'[16] are given below; other organisations which subsequently laid down procedures for the development of methods of analysis in their particular sector followed these principles to a significant degree. They require that preference should be given to methods of analysis the reliability of which have been established in respect of the following criteria, selected as appropriate:

- specificity
- accuracy

- precision; repeatability intra-laboratory (within laboratory), reproducibility inter-laboratory (within laboratory and between laboratories)
- limit of detection
- sensitivity
- practicability and applicability under normal laboratory conditions
- other criteria which may be selected as required.

8.6.2 The European Union

The Union is attempting to harmonise sampling and analysis procedures in an attempt to meet the current demands of the national and international enforcement agencies and the likely increased problems that the open market will bring. To aid this the Union issued a Directive on Sampling and Methods of Analysis.[7] The Directive contains a technical annex, in which the need to carry out a collaborative trial before it can be adopted by the Community is emphasised.

The criteria to which Community methods of analysis for foodstuffs should now conform are as stringent as those recommended by any international organisation following adoption of the Directive. The requirements follow those described for Codex above, and are given in the Annex to the Directive. They are:

1. Methods of analysis which are to be considered for adoption under the provisions of the Directive shall be examined with respect to the following criteria:
   (i) specificity
   (ii) accuracy
   (iii) precision; repeatability intra-laboratory (within laboratory), reproducibility inter-laboratory (within laboratory and between laboratories)
   (iv) limit of detection
   (v) sensitivity
   (vi) practicability and applicability under normal laboratory conditions
   (vii) other criteria which may be selected as required.
2. The precision values referred to in 1 (iii) shall be obtained from a collaborative trial which has been conducted in accordance with an internationally recognised protocol on collaborative trials (e.g. International Organisation of Standardization 'Precision of Test Methods').[17] The repeatability and reproducibility values shall be expressed in an internationally recognised form (e.g. the 95% confidence intervals as defined by ISO 5725/1981). The results from the collaborative trial shall be published or be freely available.
3. Methods of analysis which are applicable uniformly to various groups of commodities should be given preference over methods which apply to individual commodities.

4. Methods of analysis adopted under this Directive should be edited in the standard layout for methods of analysis recommended by the International Organisations for Standardization.

8.6.3 Other organisations

There are other international standardising organisations, most notably the European Committee for Standardization (CEN) and AOACI, which follow similar requirements. Although CEN methods are not prescribed by legislation, the European Commission places considerable importance on the work that CEN carries out in the development of specific methods in the food sector; CEN has been given direct mandates by the Commission to publish particular methods, e.g. those for the detection of food irradiation. Because of this some of the methods in the food sector being developed by CEN are described below. CEN, like the other organisations described above, has adopted a set of guidelines to which its Methods Technical Committees should conform when developing a method of analysis. The guidelines are:

> Details of the interlaboratory test on the precision of the method are to be summarised in an annex to the method. It is to be stated that the values derived from the interlaboratory test may not be applicable to analyte concentration ranges and matrices other than given in annex.
>
> The precision clauses shall be worded as follows:
>
> Repeatability: The absolute difference between two single test results found on identical test materials by one operator using the same apparatus within the shortest feasible time interval will exceed the repeatability value r in not more than 5% of the cases.
>
> The value(s) is (are): …
>
> Reproducibility: The absolute difference between two single test results on identical test material reported by two laboratories will exceed the reproducibility, R, in not more than 5% of the cases.
>
> The value(s) is (are): …
>
> There shall be minimum requirements regarding the information to be given in an Informative Annex, this being:
>
> Year of interlaboratory test and reference to the test report (if available)
> Number of samples
> Number of laboratories retained after eliminating outliers
> Number of outliers (laboratories)
> Number of accepted results
> Mean value (with the respective unit)
> Repeatability standard deviation (s_r) (with the respective unit)
> Repeatability relative standard deviation (RSD_r) (%)

Repeatability limit (r)w(with the respective units)
Reproducibility relative standard deviation (s_R) (with the respective unit)
Reproducibility relative standard deviation (RSD_R) (%)
Reproducibility (R) (with the respective unit)
Sample types clearly described
Notes if further information is to be given.

8.6.4 Requirements of official bodies

Consideration of the above requirements confirms that in future all methods must be fully validated if at all possible, i.e. have been subjected to a collaborative trial conforming to an internationally recognised protocol. In addition this, as described above, is now a legislative requirement in the food sector of the European Union. The concept of the valid analytical method in the food sector, and its requirements, is described below.

8.6.5 Requirements for valid methods of analysis

It would be simple to say that any new method should be fully tested for the criteria given above. However, the most 'difficult' of these is obtaining the accuracy and precision performance criteria.

Accuracy

Accuracy is defined as the closeness of the agreement between the result of a measurement and a true value of the measureand.[18] It may be assessed with the use of reference materials. However, in food analysis, there is a particular problem.

In many instances, though not normally for food additives and contaminants, the numerical value of a characteristic (or criterion) in a Standard is dependent on the procedures used to ascertain its value. This illustrates the need for the (sampling and) analysis provisions in a Standard to be developed at the same time as the numerical value of the characteristics in the Standard are negotiated to ensure that the characteristics are related to the methodological procedures prescribed.

Precision

Precision is defined as the closeness of agreement between independent test results obtained under prescribed conditions.[19] In a standard method the precision characteristics are obtained from a properly organised collaborative trial, i.e. a trial conforming to the requirements of an International Standard (the AOAC/ISO/IUPAC Harmonised Protocol or the ISO 5725 Standard). Because of the importance of collaborative trials, and the resource that is now being devoted to the assessment of precision characteristics of analytical methods before their acceptance, they are described in detail below.

Collaborative trials

As seen above, all 'official' methods of analysis are required to include precision data. These may be obtained by subjecting the method to a collaborative trial conforming to an internationally agreed protocol. A collaborative trial is a procedure whereby the precision of a method of analysis may be assessed and quantified. The precision of a method is usually expressed in terms of repeatability and reproducibility values. Accuracy is not the objective.

Recently there has been progress towards a universal acceptance of collaboratively tested methods and collaborative trial results and methods, no matter by whom these trials are organised. This has been aided by the publication of the IUPAC/ISO/AOAC Harmonisation Protocol on Collaborative Studies.[14] That Protocol was developed under the auspices of the International Union of Pure and Applied Chemists (IUPAC) aided by representatives from the major organisations interested in conducting collaborative studies. In particular, from the food sector, the AOAC International, the International Organisation for Standardisation (ISO), the International Dairy Federation (IDF), the Collaborative International Analytical Council for Pesticides (CIPAC), the Nordic Analytical Committee (NMKL), the Codex Committee on Methods of Analysis and Sampling and the International Office of Cocoa and Chocolate were involved. The Protocol gives a series of 11 recommendations dealing with:

- the components that make up a collaborative trial
- participants
- sample type
- sample homogeneity
- sample plan
- the method(s) to be tested
- pilot study/pre-trial
- the trial proper.

8.6.6 Statistical analysis

It is important to appreciate that the statistical significance of the results is wholly dependent on the quality of the data obtained from the trial. Data that contain obvious gross errors should be removed prior to statistical analysis. It is essential that participants inform the trial co-ordinator of any gross error that they know has occurred during the analysis and also if any deviation from the method as written has taken place. The statistical parameters calculated and the outlier tests performed are those used in the internationally agreed Protocol for the Design, Conduct and Interpretation of Collaborative Studies.[14]

8.7 Standardised methods of analysis for contaminants

There are many organisations that publish standardised methods of analysis for contaminants, such methods normally having been validated through a collaborative trial organised to conform to one of the internationally accepted protocols described previously. Such organisations will include AOACI, the European Organisation for Standardisation (CEN) and the Nordic Committee for Food Analysis (NMKL). Within Europe, the most important of these international standardising organisations is probably CEN. CEN has a technical committee dealing with horizontal methods of analysis in which both additive and contaminant methods of analysis are discussed (TC 275). The methods of analysis for contaminants within its work programme are outlined below. This is given by Working Group. The titles under the Working Group heading refer to the work item (topic area) of that Working Group. The Working Groups not listed (e.g. 1, 2, etc.) are concerned with additive methods of analysis.

Work programme of CEN TC 275 Working Group 3: Pesticides in Fatty Foods
Work Item A: determination of pesticides and polychlorinated biphenyls (PCBs):

Part 1: general considerations
Part 2: extraction of fat, pesticides and PCBs and determination of fat content
Part 3: clean-up methods
Part 4: determination, confirmatory tests, miscellaneous.

Work programme of CEN TC 275 Working Group 4: Pesticides in Non-Fatty Foods
Work Item A: multiresidue methods for the gas chromatographic determination of pesticide residues:

Part 1: general considerations
Part 2: methods for extraction and clean-up
Part 3: determination and confirmatory tests.

Work Item B: determination of dithiocarbamate and thiuram disulfide residues:

Part 1 spectrometric method
Part 2: gas chromatographic method
Part 3: xanthogenate method.

Work Item C: determination of bromide residues:

Part 1: determination of total bromide as inorganic bromide
Part 2: determination of bromide.

Work Item D: determination of N-methyl carbamate residues.

Work Item E: determination of benomyl, carbendazim, thiabendazole and thiophanate-methyl.

Work programme of CEN TC 275 Working Group 5: Biotoxins
Work Item A: determination of aflatoxin B_1 and/or the sum of B_1, B_2, G_1 and G_2 in cereals, shell fruits and derived products – high-performance liquid

chromatographic method with postcolumn derivatisation and immunoaffinity column.

Work Item B: determination of ochratoxin A in cereals and cereal products:

Part 1: HPLC method with silica gel clean-up
Part 2: HPLC method with bicarbonate clean-up.

Work Item C: determination of ochratoxin A in cereals and cereal products – HPLC method with immunoaffinity clean-up.

Work Item D: determination of patulin content.

Work Item E: determination of fumonisins.

Work Item F: criteria of analytical methods for mycotoxins – CEN-Report.

Work Item G: determination of domoic acid in mussels.

Work Item H: determination of aflatoxin B_1 and total aflatoxins by immunoaffinity column clean-up and HPLC in fig paste, pistachios, peanut butter and paprika powder.

Work Item I: determination of okadaic acid and dinophysis toxin in mussels by HPLC.

Work Item J: determination of saxitoxin and dicarbamoyl saxitoxin in mussels by HPLC.

Work Item K: determination of aflatoxin M1 in liquid milk.

Work programme of CEN TC 275 Working Group 6: Microbiology
Work Item A: enumeration of *Staphylococcus aureus*:

Part 1: colony count technique with confirmation of colonies (ISO/DIS 6888-1: 1997)
Part 2: colony count technique without confirmation of colonies (ISO/DIS 6888-2: 1997).

Work Item B: horizontal method for the detection of coagulase positive *Staphylococci* (*Staphylococci aureus* and other species).

Work Item C: horizontal method for the detection and enumeration of *Listeria monocytogenes*:
Part 1: detection method
Part 2: enumeration method.

Work Item D: enumeration of *Clostridium perfringens* – colony count technique.

Work Item E: horizontal method for the detection of *Salmonella*.

Work Item F: general guidance for enumeration of *Bacillus cereus* – colony count technique at 30°C.

Work Item G: detection of thermotolerant *Campylobacter*.

Work Item H: detection of *Yersinia enterocolitica*.

Work Item I: preparation of the test sample, of initial suspension and of decimal dilutions, for microbiological examination:

Part 1: general rules for the preparation of the initial suspension and of decimal dilutions
Part 2: specific rules for the preparation of the test samples and initial suspension of meat and meat products
Part 3: specific rules for the preparation of the test samples and initial suspension of milk and milk products
Part 4: specific rules for the preparation of the test samples and initial suspension of fish products
Part 5: specific rules for the preparation of the test samples and initial suspension of products other than milk and milk products, meat and meat products and fish products.

Work Item J: general guidance for microbiological examinations.

Work Item K: validation of alternative microbiological methods.

Work Item L: guidelines on quality assurance and performance testing of culture media:

Part 1: quality assurance of culture media in the laboratory
Part 2: performance testing
Part 3: practical implementation of the general guideline on quality assurance of culture media in the laboratory.
Part 4: performance testing of culture media.

Work Item M: horizontal method for the detection of *Escherichia coli* O 157.

Work Item N: horizontal method for the enumeration of *Bacillus cereus*.

Work Item O: guidelines on quality assurance and performance testing of culture media (to be elaborated as European Prestandards C02/97, C03/97):

Part 1: quality assurance of culture media in the laboratory
Part 2: practical implementation of the general guidelines on quality assurance of culture media in the laboratory
Part 3: performance testing.

Work programme of CEN TC 275 Working Group 10: Determination of Trace Elements

Work Item A: determination of trace elements – general considerations.

Work Item B: determination of mercury by CVAAS after pressure digestion.

Work Item C: determination of lead and cadmium by ETAAS after dry ashing.

Work Item D: performance criteria and general considerations.

Work Item E: pressure digestion.

Work Item F: determination of lead, cadmium, chromium and molybdenum by ETAAS after pressure digestion.

Work Item G: determination of lead, cadmium, zinc, copper, iron, chromium and nickel after dry ashing.

Work Item H: determination of lead and cadmium by ETAAS after microwave digestion.

Work programme of CEN TC 275 Working Group 11: Genetically Modified Organisms

Work Item A: detection of genetically modified organisms and derived products – sampling.

Work Item B: detection of genetically modified organisms and derived products – nucleic acid extraction.

Work Item C: detection of genetically modified organisms and derived products – qualitative nucleic acid based methods.

Work Item D: detection of genetically modified organisms and derived products – protein-based methods.

8.8 Conclusion and future trends

This chapter has outlined the key components of an internal quality control (IOC) system for laboratories, including such elements as proficiency testing as a means of validating the reliability of laboratory data. The appendix at the end of the chapter provides more detailed information on quality assurance requirements and is designed to provide the basis for a self-audit guide or an audit questionnaire. It is hoped that the chapter will help laboratories to develop robust quality systems and those auditing them to do so more effectively.

There is current discussion on an international basis whereby the present procedure by which specific methods of analysis are incorporated into legislation are replaced by one in which method performance characteristics are specified. This is because by specifying a single method:

- the analyst is denied freedom of choice and thus may be required to use an inappropriate method in some situations
- the procedure inhibits the use of automation
- it is administratively difficult to change a method found to be unsatisfactory or inferior to another currently available.

As a result the use of an alternative approach whereby a defined set of criteria to which methods should comply without specifically endorsing specific methods is being considered and slowly adopted in some sectors of food analysis. This approach will have a considerable impact on the food analytical laboratory. There are a number of issues that are of concern to the food analytical community of which analysts should be aware. These are outlined briefly below.

8.8.1 Measurement uncertainty

It is increasingly being recognised both by laboratories and the customers of laboratories that any reported analytical result is an estimate only and the 'true value' will lie within a range around the reported result. The extent of the range for any analytical result may be derived in a number of different ways, e.g. using the results from method validation studies or determining the inherent variation through different components within the method, i.e. estimating these variances as standard deviations and developing an overall standard deviation for the method. There is some concern within the food analytical community as to the most appropriate way to estimate this variability.

8.8.2 In-house method validation

There is concern in the food analytical community that although methods should ideally be validated by a collaborative trial, this is not always feasible for economic or practical reasons. As a result, IUPAC guidelines are being developed for in-house method validation to give information to analysts on the acceptable procedure in this area. These guidelines should be finalised by the end of 2001.

8.8.3 Recovery

It is possible to determine the recovery that is obtained during an analytical run. Internationally harmonised guidelines have been prepared which indicate how recovery information should be handled. This is a contentious area amongst analytical chemists because some countries of the organisations require analytical methods to be corrected for recovery, whereas others do not. Food analysts should recognise that this issue has been addressed on an international basis.

8.9 References

1. EUROPEAN UNION, *Council Directive 89/397/EEC on the Official Control of Foodstuffs*, O.J. L186 of 30.6.1989.
2. EUROPEAN UNION, *Council Directive 93/99/EEC on the Subject of Additional Measures Concerning the Official Control of Foodstuffs*, O.J. L290 of 24.11.1993.
3. EUROPEAN COMMITTEE FOR STANDARDIZATION, *General Criteria for the Operation of Testing Laboratories – European Standard EN 45001*, Brussels, CEN/CENELEC, 1989.
4. ORGANISATION FOR ECONOMIC CO-OPERATION AND DEVELOPMENT, *Decision of the Council of the OECD of 12 Mar 1981 Concerning the Mutual Acceptance of Data in the Assessment of Chemicals*, Paris, OECD, 1981.
5. EUROPEAN COMMITTEE FOR STANDARDIZATION, *General Criteria for the Assessment of Testing Laboratories – European Standard EN45002*, Brussels, CEN/CENELEC, 1989.
6. EUROPEAN COMMITTEE FOR STANDARDIZATION, *General Criteria for Laboratory Accreditation Bodies – European Standard EN45003*, Brussels, CEN/CENELEC, 1989.
7. EUROPEAN UNION, *Council Directive 85/591/EEC Concerning the Introduction of Community Methods of Sampling and Analysis for the Monitoring of Foodstuffs Intended for Human Consumption*, O.J. L372 of 31.12.1985.
8. INTERNATIONAL ORGANIZATION FOR STANDARDIZATION, *General Requirements for the Competence of Calibration and Testing Laboratories ISO/IEC 17025,* Geneva, ISO, 1999.
9. CODEX ALIMENTARIUS COMMISSION, *Report of the 21st Session of the Codex Committee on Methods of Analysis and Sampling – ALINORM 97/23A*, Rome, FAO, 1997.
10. CODEX ALIMENTARIUS COMMISSION, *Report of the 22nd Session of the Codex Alimentarius Commission – ALINORM 97/37,* Rome, FAO, 1997.
11. INTERNATIONAL UNION OF PURE AND APPLIED CHEMISTRY, *The International Harmonised Protocol for the Proficiency Testing of (Chemical) Analytical Laboratories*, ed. Thompson M and Wood R, *Pure Appl. Chem.*, 1993 65 2123–2144 (also published in *J. AOAC International*, 1993 76 926–940).
12. INTERNATIONAL UNION OF PURE AND APPLIED CHEMISTRY, *Guidelines on Internal Quality Control in Analytical Chemistry Laboratories*, ed. Thompson M and Wood R, *Pure Appl. Chem.,* 1995 67 649–666.
13. INTERNATIONAL ORGANIZATION FOR STANDARDIZATION, *Calibration and Testing Laboratory Accreditation Systems – General Requirements for Operation and Recognition – ISO/IEC Guide 58*, Geneva, ISO, 1993.
14. HORWITZ W, 'Protocol for the Design, Conduct and Interpretation of Method Performance Studies', *Pure Appl. Chem*, 1988 60 855–864 (revision published 1995).

15. HORWITZ W, 'Evaluation of Analytical Methods for Regulation of Foods and Drugs', *Anal. Chem.*, 1982 54 67A-76A.
16. CODEX ALIMENTARIUS COMMISSION, *Procedural Manual of the Codex Alimentarius Commission – Tenth Edition,* Rome, FAO, 1997.
17. INTERNATIONAL ORGANIZATION FOR STANDARDIZATION, *Precision of Test Methods – Standard 5725*, Geneva, ISO, 1981 (revised 1986 with further revision in preparation).
18. INTERNATIONAL ORGANIZATION FOR STANDARDIZATION, *International Vocabulary for Basic and General Terms in Metrology – 2nd Edition*, Geneva, ISO, 1993.
19. INTERNATIONAL ORGANIZATION FOR STANDARDIZATION, *Terms and Definitions used in Connection with Reference Materials – ISO Guide 30*, Geneva, ISO, 1992.

Appendix: Information for potential contractors on the analytical quality assurance requirements for food chemical surveillance exercises

Introduction

The FSA undertakes surveillance exercises, the data for which are acquired from analytical determinations. The Agency will take measures to ensure that the analytical data produced by contractors are sufficient with respect to analytical quality, i.e. that the results obtained meet predetermined analytical quality requirements such as fitness-for-purpose, accuracy and reliability. Thus when inviting tenders FSA will ask potential contractors to provide information regarding the performance requirements of the methods to be used in the exercise, e.g. limit of detection, accuracy, precision, etc., and the quality assurance measures used in their laboratories. When presenting tenders laboratories should confirm how they comply with these specifications and give the principles of the methods to be used. These requirements extend both to the laboratory as a whole and to the specific analytical determinations being required in the surveillance exercise. The requirements are described in three parts, namely:

- Part A: Quality assurance requirements for surveillance projects provided by potential contractors at the time ROAMEs are completed and when commissioning a survey
- Part B: Information to be defined by the FSA customer once the contract has been awarded – to be agreed with contractor
- Part C: Information to be provided by the contractor on an on-going basis once contract is awarded – to be agreed with the customer.

Each of these considerations is addressed in detail below. Potential contractors are asked to provide the information requested in Part A of this document when submitting ROAME forms in order to aid the assessment of the relative merits of each project from the analytical/data quality point of view. This information is best supplied in tabular form, for example that outlined in Part A, but may be provided in another format if thought appropriate. The tables should be expanded as necessary. Parts B and C should not be completed when submitting completed ROAME forms.

Explanation of Parts A, B and C of Document

Part A

Part A describes the information that is to be provided by potential contractors at the time that the ROAME Bs are completed for submission to the Group. Provision of this information will permit any FSA 'Analytical Group' and customers to make an informed assessment and comparison of the analytical quality of the results that will be obtained from the potential contractors bidding for the project. Previously potential contractors have not been given defined guidance on the analytical quality assurance information required of them and this has made comparison between potential contractors difficult. Part A is supplied to potential contractors at the same time as further information about the project is supplied.

The list has been constructed on the premise that contractors will use methods of analysis that are appropriate and accredited by a third party (normally UKAS), participate in and achieve satisfactory results in proficiency testing schemes and use formal internal quality control procedures. In addition, Parts B and C are made available to the potential contractors so that they are aware of what other demands will be made of them and can build the costs of providing the information into their bids.

Part B

This section defines the analytical considerations that must be addressed by both the customer and contractor before the exercise commences. Not all aspects may be relevant for all surveys, but each should be considered for relevancy. Agreement will signify a considerable understanding of both the analytical quality required and the significance of the results obtained.

Part C

This section outlines the information that must be provided by the contractor to a customer on an on-going basis throughout the project. The most critical aspect is the provision of Internal Quality Control (IQC) control charts thus ensuring that the customer has confidence that the contractor is in 'analytical control'.

By following the above the FSA customers will have confidence that the systems are in place in contractors with respect to analytical control and that they are being respected. It is appreciated that not all aspects outlined in Parts A, B and C will be appropriate for every contract but all should be at least considered as to their appropriateness.

Contents of Parts A, B and C of document

Part A

Potential contractors should provide the information requested below. Please provide the information requested either in section 1 or in section 2 and then that in sections 3 to 5.

Section 1: Formal quality system in the laboratory if third party assessed (i.e. if UKAS accredited or GLP compliant)

Please describe the quality system in your laboratory by addressing the following aspects:

- To which scheme is your laboratory accredited or GLP compliant?
- Please describe the scope of accreditation, by addressing:
    1. the area that is accredited
    2. for which matrices, and
    3. for which analytes

  or supply a copy of your accreditation schedule.
- Do you foresee any situation whereby you will lose accreditation status due to matters outside your immediate control, e.g. closure of the laboratory?

Section 2: Quality system if not accredited

Please describe the quality system in your laboratory by addressing the following aspects:

- Laboratory Organisation:
    1. Management/supervision
    2. Structure and organisational chart
    3. Job descriptions if appropriate.
- Staff:
    1. Qualifications
    2. Training records
    3. Monitoring of the analytical competency of individual staff members.
- Documentation
    1. General lab procedures
    2. Methods to be used (adequate/detailed enough to control consistent approach).
- Sample Preparation
    1. Location
    2. Documented procedures
    3. Homogenisation
    4. Sub-sampling
    5. Sample identification
    6. Cross-contamination risk
    7. Special requirements.
- Equipment Calibration
    1. Frequency

2. Who
3. Records
4. Marking.

- Traceability
  1. Who did what/when
  2. Equipment – balances, etc.
  3. Sample storage/temperature
  4. Calibration solutions: how prepared and stored.
- Results/Reports
  1. Calculation checks
  2. Typographic checks
  3. Security/confidentiality of data
  4. Software usage/control
  5. Job title of approved signatory.
- Laboratory Information management System

Please outline the system employed.

- Internal Audits
  1. Audit plan
  2. Frequency
  3. Who carries out the audit?
  4. Are internal audit reports available?
  5. What are the non-compliance follow-up procedures?
- Sub-contracting
  1. In what circumstances is sub-contracting carried out?
  2. How is such sub-contracting controlled and audited?

Section 3: Proficiency testing

Please describe the arrangements for external proficiency testing in your laboratory by addressing the following aspects:

- Do you participate in proficiency testing schemes? If so, which schemes?
- Which analyte/matrices of the above schemes do you participate in?
- What are your individual proficiency scores and their classification, (e.g. z-scores or equivalent), over the past two years, for the analyte/matrices of relevance to this proposal?
- What remedial action do you take if you should get unsatisfactory results?

Section 4: Internal quality control

Please describe the IQC measures adopted in your laboratory by addressing the following:

- What control samples do you use in an analytical run?
- Do you follow the Harmonised Guidelines?[1]

[1] 'Guidelines on Internal Quality Control in Analytical Chemistry Laboratories', ed. M. Thompson and R. Wood, *Pure Appl. Chem.*, 1995, **67**, 649–66.

- What IQC procedures are in place?
- Do you use Certified Reference Materials (CRMs), and if so, how? For example, specify the concentration(s) matrix type(s), etc.
- Which appropriate CRMs do you use?
- Do you use In-House Reference Materials (IHRM) and how are they obtained? For example, specify the concentration(s) matrix type(s).
- Are they traceable? For example, to CRM, a reference method, inter-laboratory comparison, or other.
- What criteria do you have regarding reagent blanks?
- What action/warning limits are applied for control charts?
- What action do you take if the limits are exceeded?
- Do you check new control materials and calibration standards? If so, how?
- Can we see the audit of previous results – what actions have been taken or trends observed?
- Do you make use of duplicate data as an IQC procedure?
- How frequently are control materials (CRMs, blanks, IHRM, etc.) incorporated in the analytical run?
- Do you randomise your samples in an analytical run? (including duplicates).

Section 5: Method validation
Please describe the characteristics of the method of analysis you propose to use in the survey by addressing the following:

- What methods do you have to cover the matrix and analyte combinations required?
- Do you routinely use the method?
- Is the method accredited?
- Has the method been validated by collaborative trial (i.e. externally)?
- Has the method been validated through any In-House Protocol?
- Is it a Standard (i.e. published in the literature or by a Standards Organisation) Method?
- Please identify the performance characteristics of the methods, i.e.
    1. LOQ
    2. LOD
    3. Blanks
    4. Precision values over the relevant concentration range expressed as relative standard deviations
    5. Bias and recovery characteristics including relevant information on traceability.
- Do you estimate measurement uncertainty/reliability?
- Do you normally give a measurement uncertainty/reliability when reporting results to your customer?

Part B
The FSA customer is to consider and then define the following in consultation with the contractor:

1. What analysis is required for what matrices.
2. The sample storage conditions to be used. Are stability checks for specific analytes undertaken?
3. The methods to be used and a copy of Standard Operating Procedures (SOPs) where accredited, including any sampling and sample preparation protocols, to be supplied to the customer.
4. The IQC procedures to be used. In particular the following should be considered:
   - the use of the International Harmonised Guidelines for IQC
   - the use of control charts
   - randomisation within the run
   - the composition of the analytical run (e.g. the number of control samples, and in particular the number of blanks, spikes, IHRMs, etc.)
   - the reference materials to be used
   - the determination of recoveries on each batch using procedures as described in the International Harmonised Guidelines with all results to be corrected for recovery except where otherwise specified (i.e. for pesticides) and for the recovery data quoted to be reported.
5. The measurement limits (i.e. limit of detection (LOD): limit of determination/quantification (LOQ), and reporting limits, etc.).
6. The maximum acceptable measurement reliability (also known as measurement uncertainty) for each analytical result.
7. The treatment of individual results with respect to uncertainty, reliability, i.e. as
   (a) $x \pm y$ μg/kg where y is the measurement reliability (i.e. as if the sample were to be a 'historic' surveillance result), or
   (b) not less/more than x μg/kg where x is the analytical result determined less the measurement reliability (i.e. as if the sample were to be an 'enforcement style' result) when assessing compliance with a (maximum) limit.
8. Whether there are to be action limits whereby the customer is immediately notified of 'abnormal' results.
9. The procedures to be used for confirmation of 'abnormal' results, e.g. those that exceed any defined statutory limit. The procedures to be used if qualitative analysis is to be undertaken.
10. The consistent way of expressing results, e.g. (a) on a wet (as is) basis, on a dry weight basis or on a fat weight basis, and (b) the reporting units for specific analytes to be used throughout survey (i.e. mg/kg, etc.).
11. The time interval for customer visits (e.g. once every three months, or as otherwise appropriate) and for submission of control charts.
12. Whether there are any possibilities of developing integrated databases between customer and major contractors. If not, the customer to provide reporting guidelines.
13. The procedure for logging in of samples and traceability of sample in the laboratory.
14. The security of samples within the laboratory.

Part C

The following are to be provided by the contractor on an on-going basis throughout the contract to confirm that the contractor remains in 'analytical control'.

1. Copies of the control charts and duplicate value control charts or other agreed measures to monitor IQC.
2. Records of action taken to remedy out-of-control situations to be provided at the same time with control charts.
3. Where action limits have been identified in Part B (see para. 8), the results of samples that exceed the action limits are to be sent to the customer as soon as available.
4. Any relevant proficiency testing scheme results obtained during the course of the survey.

Part III

Other types of audit

9

Benchmarking

D. Adebanjo, Leatherhead Food Research Association

9.1 Introduction

Benchmarking has been defined in various ways by different companies and practitioners. Rank Xerox, a pioneer user of the technique, has a working definition as follows:

> Benchmarking is the search for industry best practices that lead to superior performance.

The benchmarking process thus involves comparison in order to identify, understand and adapt superior practices. Comparison may be either within the same organisation or with an external organisation. Subsequently, the modification or change of current practices or methodologies is intended to lead to improvements in overall business effectiveness. The overall aim of benchmarking is to provide a structured and rational process to promote and support continuous improvement.

9.1.1 History of benchmarking

Rank Xerox, part of Xerox Corp, is often credited with the pioneering role in developing benchmarking as an improvement technique. From the mid-1960s to the mid-1970s, its profits rose about 20 per cent per annum but, by the late 1970s, the company was losing a significant share of the photocopier market to its Japanese competitors. In 1979, Xerox started competitive benchmarking. This initially involved analyses of unit manufacturing costs and comparison of operating capabilities and features with competing products.

Xerox then sought the support of its Japanese affiliate, Fuji-Xerox, in more formalised analysis. Investigations revealed that Xerox's Japanese competitors could sell copiers cheaper than the Americans could produce them. Consequently, Xerox set new cost targets and developed a benchmarking process to ensure that improvements were made. The net effect was that Xerox was eventually able to win back some of its market share from the Japanese.

Xerox's success with benchmarking led to wide acceptance of the concept in America in the 1980s. This was also fuelled by a realisation of the need to improve business performance and the growing adoption of Total Quality Management (TQM) principles. Companies such as AT&T, General Electric, Milliken, General Motors, Motorola and DuPont all applied benchmarking to their operations.

Benchmarking was introduced to the UK in the late 1980s and early 1990s. UK companies with American connections, such as Milliken Industrials and Rank Xerox, were primarily responsible for the initial impetus. Other companies, such as Royal Mail, Rover Group, ICL and British Telecom, quickly identified the importance of benchmarking as an improvement tool and applied it to their businesses. Benchmarking is now widely used in the UK and has enjoyed active support from government departments including the Department of Trade and Industry (DTI), Ministry of Agriculture, Fisheries and Food (MAFF) and the Department for Education and Employment (DfEE).

9.1.2 Benefits of benchmarking

The ever-increasing pace of change and competitiveness in the marketplace imposes a need to maintain or even increase competitive advantage. Organisations must tailor business tools and methods to meet their specific developmental needs. By focusing on the external environment and improving process efficiency, benchmarking promotes a climate for improvement-oriented change. The most significant benefits of benchmarking are:

- Establishing performance goals and objectives – by focusing on the external environment, benchmarking confirms the need and extent of change required.
- Increasing efficiency by improving processes – understanding the organisation's work practices and improving them supports improvement of overall business performance.
- Promoting breakthrough thinking – by encouraging staff to think 'outside the box'.
- Providing a vision of excellence – by understanding practices that have been successfully used elsewhere and which are superior to current internal practices.
- Measuring productivity and managing change – by understanding process outputs and monitoring them through the change management process.
- Improving competitiveness – by providing a better appreciation of markets and products and challenging current business methods.

- Promoting cultural change – by allowing employees to focus on continuous improvement and how they compare with others.

9.1.3 Benchmarking pitfalls

Prior to starting benchmarking, it is important to understand some of the common misconceptions about the technique. It is worth remembering that benchmarking is a structured and rational process and not:

- the practice of cloning
- measuring to compile league tables
- industrial visits
- a series of 'wow' trips to keep up with the Jones's
- inapplicable to our business.

In addition to these misconceptions, failures in benchmarking have been attributed to the following:

- lack of management commitment
- lack of focus and perseverance
- lack of communication
- expecting 'quick fix' solutions
- failure to understand an organisation's own processes as the starting point for benchmarking
- comparing with the wrong kind of process or organisation
- failure to measure the right indicators
- failure to implement improvements and link benchmarking to the organisation's overall strategy for improvement.

9.2 Basic principles of benchmarking

9.2.1 Introduction

The widespread use of benchmarking has led to different practitioners developing different process steps to benchmarking. It is not unusual to come across the 6-, 10- or 12-step benchmarking process. Irrespective of the differences in process steps, there is a high level of uniformity of the basic concepts underlying the practical application of benchmarking. Broadly speaking, the different process steps support four phases (or stages) of benchmarking. These are:

1. Planning phase – identification of business process or function to be benchmarked, external partners and how benchmarking would be carried out.
2. Analysis phase – actual collection of data and analysis for performance gaps.
3. Action phase – communication of findings, setting of targets and implementation of specific improvement actions.

4. Review phase – identifying learning points, evaluating the benefits of the process and continuous monitoring of improvement.

9.2.2 Benchmarking process

Figure 9.1 provides an overview of the basic components of the 6-, 10- and 12-step processes. The main difference appears to be the level of detail favoured by the originators. The success of benchmarking will probably depend more on the correct application of the chosen model as opposed to the characteristics of the model itself. A detailed discussion of each of these steps is beyond the scope of this chapter, but sections 9.4 to 9.6 give some insight into the key issues concerned.

9.2.3 Types of benchmarking

It is generally recognised that there are four distinct types of benchmarking that can be conducted by an organisation:

1. internal benchmarking
2. competitive benchmarking
3. functional benchmarking, and
4. generic (or best practice) benchmarking.

Each of these has different strengths and weaknesses that, consequently, make them applicable to different organisations and circumstances.

Internal benchmarking

This refers to benchmarking within the same organisation. Typically, within multinational or multidivisional firms, there are similar functions at different locations or operating units. Comparison of these internal functions constitutes internal benchmarking. This may involve, for example, comparison of manufacturing or distribution operations among different departments or among US, UK and Japanese operations.

Internal benchmarking is the easiest form of benchmarking. Co-operation and data are readily available and concerns about confidentiality and trust are easily overcome. This form of benchmarking also promotes sharing and communication within the organisation. However, the benchmarking improvements can only be as good as the best division within the organisation, and major breakthroughs are unlikely.

Very often, organisations adopt internal benchmarking as a starting point for process development. Once the company has gained adequate returns and is comfortable with the benchmarking process, it is usual to seek external benchmarking partners. TNT (UK) Ltd is a good example of a company that adopted this approach to benchmarking.

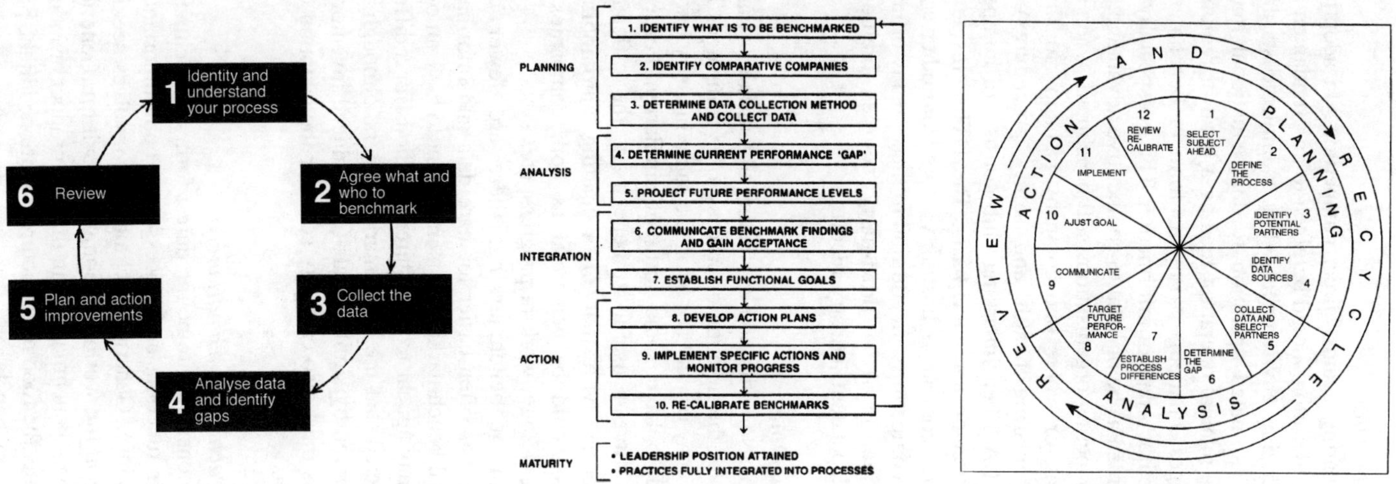

Fig. 9.1 The 6-, 10- and 12-stage models of the benchmarking process. (Sources: Cook (1996), Camp (1989) and Codling (1992).)

Competitive benchmarking

This involves comparing against direct competitors. In adopting this form of benchmarking, consideration should be given to any fundamental differences in operations and operating environments. For example, costs and supply chain operations for an automated operation feeding a national market may differ significantly from a manual operation feeding a local or regional market.

Perhaps the biggest challenge with this form of benchmarking is gaining access to the competitor's data or information. This is especially true where such information is regarded as being vital to competitive advantage. In spite of these concerns, many competing organisations still benchmark when there are mutual lessons to be learnt, or where one company is happy to share some of its practices (e.g. comparing health and safety procedures). For example, Leatherhead Food RA's benchmarking club consists of competing companies that share information on a regular basis. The DTI's 'Inside UK Enterprise' involves best practice companies hosting visitors from other companies, which sometimes include competitors. In addition to these, many organisations collect competitive information from third-party sources including customers, consultants and independent industry reports.

Functional benchmarking

This involves comparison with non-competing organisations that are recognised as industry leaders for the function or process to be compared. The organisation to be benchmarked may or may not be in the same industry but the functions to be compared need to have some form of similarity. For example, a food manufacturer may find it useful to benchmark the efficiency of its lean manufacturing process with that of a leading automobile or aerospace manufacturer. This would be beneficial as both companies will have raw material deliveries, storage, work-in-progress, etc.

With this form of benchmarking, there are fewer concerns about confidentiality and it is often easier to share data and information. To benefit fully from functional benchmarking, it is important to have an open mind and be receptive to what may appear to be different methods for a different industry. It should be remembered that the benchmark partners, although from a different industry, may be driven by similar goals (e.g. reduced inventory, higher yield). Benchmark partners must also carefully consider the adoption of best practices in their organisation.

Generic (or best practice) benchmarking

This is sometimes considered to be the purest form of benchmarking. The basis of this is that certain functions and processes are both important and universally applied (e.g. invoicing). Generic benchmarking involves seeking out the best company anywhere in the world and comparing with it. Generic benchmarking differs from functional benchmarking in that the product or industry may not be limiting. For example, the food manufacturer mentioned in the previous example would not be able to benchmark lean manufacturing operations against an electricity supply company as such a company will not have manufacturing operations. It would be possible, however, to benchmark a more generic process such as recruitment, invoicing, handling of customer complaints, etc. These are processes that are carried out widely by almost all organisations.

A competent understanding of the relevant process is important for generic benchmarking. The main advantage of this form of benchmarking is that there are potentially huge gains to be made since learning is from an organisation that has already demonstrated the process to work. The difficulty with generic benchmarking is identifying which organisation is 'best'. To some extent, this will be influenced by what is most important to the seeking partner. For example, the company with the fastest process may not necessarily be the cheapest or the most efficient. Thus a company which is best practice for one organisation may not be for another organisation although it is also possible for one company to be best practice partner for many organisations.

In order to determine what type of benchmarking to use, an organisation must carefully consider its circumstances and its objectives. It is also important to be familiar and comfortable with the benchmarking process. In this respect it may be advisable to attempt some internal benchmarking, where possible, as a starting point. When experience has been gained, benchmarking may then be given an external, and possibly less friendly, dimension.

9.3 Understanding your organisation and its processes

Understanding your organisation is vital to benchmarking. Many benchmarking practitioners have devised different ways of planning for benchmarking by a review of their organisations. Some of the issues involved are discussed in this section.

A process can be defined as a sequence of steps which adds value by producing a predetermined output from a variety of inputs. It is important to remember that not all processes lead to products that are visible or apparent. However, some of these processes (e.g. logistics, invoicing) are vital to the business success of the organisation.

It will be difficult to compare business processes with internal or external benchmarking partners if these processes are not thoroughly understood. While this may sound obvious, it is worth remembering that many benchmarking initiatives fail as a result of poor preparation. In particular it is important that the process is understood for its logic and flow and that the strengths and weaknesses are identified. The process owner should also be involved, not only in the understanding of the process, but at all phases of the benchmarking process.

9.3.1 Methods for identifying processes for benchmarking

A number of ways of identifying processes are described below. These may be used individually or they may also be combined:

- *Strategic evaluation.* The broad objective is to identify strategic objectives and subsequently determine the functions and processes that underpin them. The strategic objectives may result from company needs (e.g. low level inventory), mission statements, statutory obligations (e.g. waste or emission levels), etc. They may also be determined more formally by means of a strategic assessment of the organisation. Self-assessment against an excellence model or framework provides a good opportunity for this. Self-assessment not only helps identify deliverables but also quantifies the organisation's performance. This quantification often helps determine what aspects or functions of the organisation would most benefit from improvement. Figure 9.2 shows the EFQM Excellence Model, which is widely used for self-assessment throughout Europe.
- *Questioning.* This involves asking a series of questions, often related to business and customer objectives, in order to arrive at a functional or subject area for benchmarking. Generally speaking, the subject areas selected should be of strategic and business importance to the organisation. Questions that may be asked include:
  - What is our business about?
  - What is most critical to business success?
  - What factors would impact most on our customer/supplier relationships?
  - What functions or areas of our business are underperforming?
  - What problems have been identified in these functions?
  - What areas, when improved, will have the greatest impact on our business results?
  - In what areas of business are our competitors outperforming us?

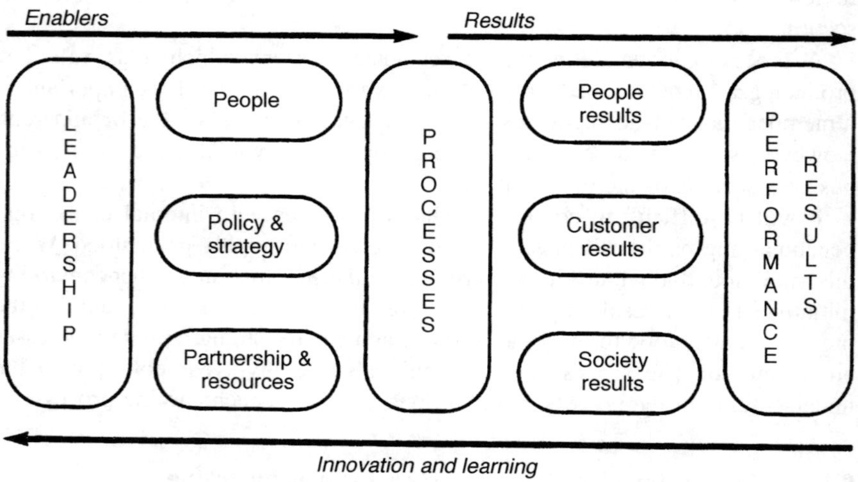

Fig. 9.2 The EFQM Excellence Model. (Source: © 1999 EFQM. The Model is a registered trademark of the EFQM (European Foundation for Quality Management).)

- *Use a team approach.* Xerox identifies a team to apply an Analytical Hierarchy Process to identify areas to benchmark. This involves weighting and ranking the alternatives by taking into account decision criteria such as resources required, ease of finding benchmarking partners, etc.
- *Brainstorming.* This can be a powerful tool in identifying potential benchmarking areas. It is always helpful to get as much buy-in from different aspects of the business as possible. Questions that may be considered during such an exercise could include the following:

  - What are the most problematic areas?
  - What must we do to achieve business success?
  - What factors are most critical to our business?
  - What can we do to improve our stakeholder perceptions/relationships?
  - Where do we face the greatest competitive pressures?
  - What are our most important resources/costs?
  - What factors impact most on our profitability?
  - What performance measures have we applied?

- *Performance measurement.* All organisations make use of performance measures in their reporting systems. These may, on their own, indicate areas for benchmarking or otherwise act as a basis for developing other measures. It is, however, important that the current performance measures are wide-ranging and do not lean heavily towards financial results. In this respect, the organisation may consider the use of the balanced scorecard, which splits organisational performance into four different perspectives. Figure 9.3 illustrates this.
- *Offshoot from other initiatives.* Other initiatives such as customer complaints management, strategy development and problem solving may highlight subject areas that need significant improvement and that may benefit from benchmarking.

9.3.2 Detailed analysis

The identification of the broad subject area is followed by cascading downwards until the processes are at a level where they can be easily represented, evaluated and benchmarked. 'Single' processes are usually a combination of activities that may be complex and may become secondary and tertiary processes. For example, if we take serving a customer in a restaurant as a single process, this will involve a sequence of operations such as seating the customer, taking his/her order, passing the order to the kitchen, delivering the food to the table, etc. Each of these steps is not as simple as it initially appears – seating the customer may involve finding a table in a non-smoking area, ensuring the table is clean, making sure there are enough chairs for the group (perhaps including baby chairs), etc.

The process steps need to be identified, sequenced, described and recorded as fully as possible. Furthermore, the critical measurements to be used for the

Financial perspective	
Goals	Measures

Customer perspective	
Goals	Measures

Internal business perspective	
Goals	Measures

Innovation and learning perspective	
Goals	Measures

Fig. 9.3 The balanced scorecard.

benchmarking exercise need to be identified. This involves rationalisation to determine where measurement is appropriate, the units of measure to be used and the ability of the measures to reflect the process performance accurately.

These activities will require the input of all members of the team, including those who are not process owners. However, it is important that at least one member of the team works on the process and has a thorough understanding of the sequence of activities involved. This often reduces or eliminates the need to spend time on the shopfloor at the initial stages of the process analysis. However, shopfloor visits are advised in order to verify the result of the analysis or where there are significant differences between the 'theoretical' process and the 'actual' process.

Ishikawa cause-and-effect diagrams, workflow diagrams and other widely used analytical tools are commonly used in the detailed analysis of the process(es) to be benchmarked. Figures 9.4 and 9.5 show examples of the Ishikawa (or fishbone) diagram and the workflow diagram (or process map). To use the Ishikawa diagram, the following steps should be followed:

- Write the name of the primary process at the head of the fish.
- The name of each sub-process or major activity that impacts on the primary process becomes a label for one of the bones.
- Smaller bones can then be drawn from the main bones to represent activities that occur in the sub-process.

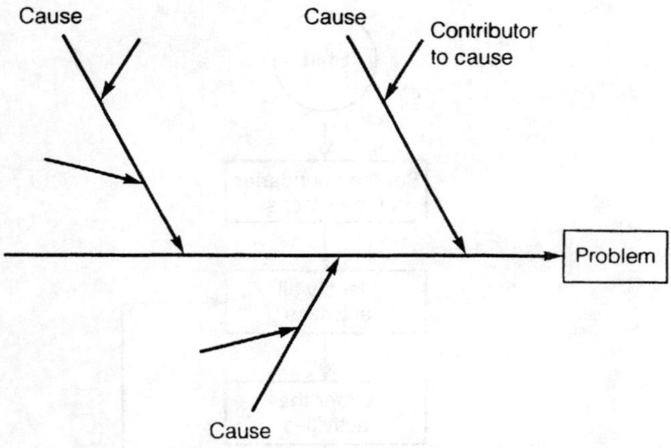

Fig. 9.4 Structure of the Ishikawa cause-and-effect diagram.

The workflow diagram consists of four symbols:

1. a start symbol (circle) to indicate the beginning of the process
2. an activity symbol (rectangle) to indicate some form of action taking place
3. a decision symbol (diamond) to indicate a point where an inspection needs to take place and a decision needs to be made
4. an end point symbol (ellipse) to indicate the completion of the process.

The use of these tools in themselves can lead to improvements even before the benchmarking process is completed. As a final check, it is advisable to re-evaluate how the chosen process relates and contributes to the identified strategic issues and, ultimately, the business objective(s) of the organisation.

9.4 Identifying potential benchmarking partners

Successful benchmarking depends heavily on the suitability of the participating partner. Not only must care be taken to ensure that the organisation is more advanced, but the processes or subject areas must be truly comparable. To put it simply, do not compare apples and pears.

Irrespective of the type of benchmarking the company has decided to adopt, the underlying principle remains the same – seek a partner in a better/best-performing section, division or company. Potential benchmarking partners may be identified from a range of information sources. These include:

- internal sources such as internal publications, databases, employee knowledge of the industry and market, contact with other organisations, e.g. at exhibitions, customer suggestions or trade meetings
- external sources such as trade journals, independent consultants, newsletters, external libraries (including universities and research associations), industry

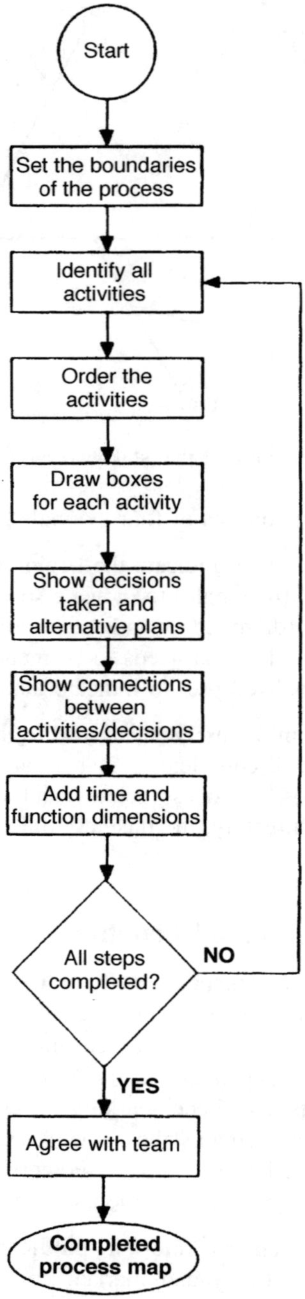

Fig. 9.5 Structure of the workflow diagram.

reviews and publications, data collection organisations and professional/trade associations
- secondary sources such as focus groups, internal/external surveys, etc.
- the DTI's 'Inside Enterprise' programme, which promotes visits to best practice organisations, thereby providing first-hand information about potential benchmarking partners
- Leatherhead Food RA, through its management of the Benchmarking Club and other contacts, which can also help identify potential benchmarking partners.

Typically, a fairly comprehensive list is drawn up initially and then pruned by identifying and applying partner selection attributes or criteria. Top level criteria may include factors such as product attributes, company size, market sector, or rate of growth. Once a preliminary shortlist is drawn up, it may be possible to apply more specific criteria, which may devolve to process level considerations. For example: which company is best at minimising product waste? What division has the most comprehensive training programme?

9.4.1 Making an approach

Making an improper approach may result in the request for a visit being denied. Some ways of making successful approaches are outlined as follows:

- *Existing relationships*. If the potential benchmarking partner is a customer or supplier, then sales or purchasing managers may be able to make initial contact to identify the most appropriate individual to discuss benchmarking with. This may also be applied if the potential partner has a relationship with a sister company or if there has been migration of personnel between the two organisations.
- *Professional contact*. Where there are no existing relationships, direct contact with an opposite number in the target organisation may be pursued. It may be possible to identify the appropriate individual from membership directories of professional associations. Many organisations that are heavily involved in benchmarking tend to have a dedicated benchmarking manager. Initial contact may be either written or verbal – the vital thing is to define clearly the objectives and mutual advantages of the proposed exercise.
- *Third-party contact*. External organisations such as the DTI programme, Leatherhead Food RA and consultants may be approached to make the initial contact with the potential partner.

9.5 Preparing for a benchmarking visit

Benchmarking is typically carried out by a team. To be effective, the team needs to be balanced and possess all necessary teamworking and functional skills. While it may be argued that there is no such thing as the ideal team composition, it is suggested that a benchmarking team consist of the following:

- A team leader who may or may not be the process owner.
- Members with a combination of the following:
  - analytical skills
  - process documentation skills
  - information search and analysis capability.

Furthermore, when needed, special expertise/knowledge by way of an internal or external customer who makes use of the process may be drafted to the team as required.

Site visits are not the only way of conducting a benchmarking study. However, they have the ability to provide the most useful insights. They are also generally viewed as being more credible than other alternatives. A successful benchmarking visit requires meticulous preparation and planning. It is necessary to get as much information about the partner as is available. In addition, the following 'rules' should be noted:

- Agree and confirm the meeting date and venue.
- Agree who should attend.
- Discuss and agree the purpose of the meeting as well as the requirements and expectations of both sides.
- Confirm the itinerary of the visit, especially if a tour of the operations is planned.
- Prepare background documentation (e.g. process specifications, questionnaire, checklist).
- Determine and prioritise questions to be asked or issues to be raised.
- Agree code of conduct and, if required, sign a confidentiality agreement.

9.5.1 Site visit

Where a benchmarking visit includes a facility tour, it may be advisable (subject to the agreed itinerary) to make this the first activity. This not only helps to serve as an ice-breaker but also gives the visiting party a clearer overview of the operations. This will not only enhance the quality of subsequent discussions but may generate some entirely new questions. When conducting a facility tour, it is advisable to speak to the appropriate operators/supervisors. Observation as well as listening are vital to getting maximum benefit from site visits.

A facility tour should then be followed up by the meeting proper. This generally takes up the greater part of the day and gives both sides the opportunity to take notes and ask questions. Once again, conduct is vital and the following should be considered during the meeting:

- Focus on the 'how'.
- Do not succumb to information 'grazing'.
- Do not be afraid to clarify any lack of understanding.

- Look beyond the obvious for the clues to superior performance.
- Ensure a two-way flow of information; be open and honest.
- Take notes.
- Use a checklist to ensure that all issues are covered.
- Do not ask the kinds of questions that you would not wish to be asked.
- Confirm if there will be a follow-up or return visit.

9.5.2 After the visit

It is important that the outcome of the visit be documented as soon as possible after the meeting to ensure that no information is lost. This may typically be preceded by a debrief of the team to confirm observations and achieve consensus. The findings from the visit must then be communicated to the relevant people within the organisation. This may take the form of distributing and discussing the report or giving a formal presentation on the visit.

A thank-you letter should be sent to the hosts, as far as possible highlighting the positive aspects of the visit. If a follow-up or return visit was agreed, it is advisable to mention this and, maybe, confirm and discuss details.

Finally, the team/organisation must now determine how to make use of the information or data gathered during the visit.

9.6 Analysis and improvement

The analysis and improvement action stages of the benchmarking process focus on the internal exploitation of information and data from the external visit. The scope of this book implies that none of these issues can be fully described here; however, the most important attributes will be discussed.

9.6.1 Gap analysis

The benchmarking team requires tools both to analyse data and to identify the root causes for the 'gap'. In depicting the gap in performance, it is also important to consider the rate of improvement.

Some of the tools that can be used for the presentation of data include bar charts, histograms, scatter diagrams, pie charts, Pareto diagrams and maturity indexes. Tools that may be used for root cause analysis include flowcharts, fishbone diagrams, force-field diagrams, SWOT analysis, control charts, brainstorming, quality function deployment and affinity diagrams.

The differential in performance between the company and its benchmarking partner may be classified in one of the following ways:

- Negative gap – this implies that the partner's practices and performance are superior and a change of internal practices will be required for superior performance to be attained.

- Parity gap – the two organisations have similar levels of performance even if methods differ slightly. Effort should be made to ensure that the company does not slip but strives for a better rate of improvement.
- Positive gap – the company's performance is superior to the partner's. The focus should be on maintaining or improving performance while also searching for new benchmarking partners who may add more value.

To analyse the differences between the two organisations fully, it may be necessary to take many measures and put these measures in the context of any differences between the organisations' operations. The 'gap' or differences between process performance can then be described quantitatively, qualitatively, or using a combination of the two methods. For example, a benchmarking visit may reveal a difference in management style and attitudes and, consequently, a higher level of employee involvement and motivation. Such a qualitative gap can be verified/quantified by examining the levels of staff absenteeism, sickness, efficiency, etc.

In an effort to benchmark its supply chain operations, Rank Xerox identified leaders in integrated supply chain management (including IBM, Hewlett-Packard, Apple and Siemens) as potential partners for benchmarking. One of the important findings from the survey was the benefits of changing from a forecast-led shipping and stockpiling of warehouses to a customer order-led shipping of products. This meant that orders would be delivered straight to the customer from a central warehouse rather than from stockpiled warehouses in different locations.

9.6.2 Detailing best practice and preparing recommendations

Following from the analysis, the team should detail the best practice ideas and methods identified from the visit and subsequent analysis. Such detail may include the use of process maps. These practices should then be matched with the performance gaps determined and subsequently viewed against the background of the organisation's internal practices.

The team then needs to finalise its recommendations on the benchmarking exercise. This may include a comparison of both companies and the reasons for performance differential. Where possible, priority areas for change may be identified as well as a quantification of the possible end point after change. The team may also use the experiences of the benchmarking partner to suggest the resources needed and time that may be allowed for change.

Finally, the recommendations should be presented to the management team (or the sponsor of the benchmarking study). The team should ensure that management fully understands the reasons for change and the implications that change would have for the company's people, processes and performance.

9.6.3 Developing and implementing action plans

Management may wish to prioritise the possible changes that may be made to the organisation. Tools that may be useful in doing this include a cost/benefit analysis. Once priorities have been determined, responsibility for change management needs to be allocated. Some of the options include line management, project team and 'process champion' (especially for cross-functional development) implementation.

The party charged with implementing change must have active commitment and a strong degree of ownership. It is also important to have good process and change management skills as well as attention to detail. The implementation party needs to consider the following.

Plans and goal setting

From the outset, the goals to be attained should be set and the related measures defined. Goals should be bold and ambitious but also flexible enough to allow for adjustment if a change in circumstances occurs. Taking a look at best practice organisations or the benchmarking partner gives an insight into realistically achievable goals.

In formulating the set of actions that constitute the improvement or change plan, it is also important to determine the skills that would be needed and match these against what is currently available within the organisation. An extension of such resource planning is the specification of the financial, opportunity and time costs required to implement change. Other guidelines for improvement and change action planning include:

- Maintain a balance between short- and long-term goals.
- Divide the plan into manageable chunks.
- Specify individual action plans for people directly involved in implementation.
- Determine a monitoring system.
- Have contingency plans.

Communicating plans

The implementation team needs to communicate its findings and plans to the wider organisation. The likelihood is that not everyone within the organisation will be aware of their work. However, commitment from the whole organisation will be needed if the change programme is to succeed, especially if the organisation aims to make benchmarking a part of its overall development strategy.

Effective communication will also help reduce resistance to change as there will be a better understanding of the need for improvement, the aspired end point and how the changes will affect processes and their owners. The implementation team might also wish to communicate the following:

- Why the organisation undertook benchmarking and the people involved.
- The timeframe for change and the processes to be changed.
- The benefits of change to the organisation as well as to individuals.
- Current accomplishments and the milestones to be attained.

9.7 Review

Review is a vital step in the benchmarking process. All stages in the benchmarking study should be monitored on a regular basis, especially during the implementation of action plans (compare with milestones). Furthermore, the company needs to re-calibrate itself after systematic review.

9.7.1 Process and progress review

An overall systematic review needs to be carried out at the end of a study. To ensure that this review is not avoided, a provisional date should be set as part of the overall plan and confirmation notified to all attendees well in advance. This should primarily involve the project team, process owners and senior management. Review should consider two elements, the results of the process benchmarking and the benchmarking process itself.

Results of process benchmarking

In this form of review, the output of benchmarking is inspected to ensure that performance goals have been attained. Generally, if there had been constant monitoring and review during the study, the company would most probably have achieved its overall goal. Evidence of improvement should be presented in addition to steps taken to ensure that such improvement is, at least, maintained.

It may also be beneficial to compare the rate of change with that of competitors before suggesting whether further work needs to be done. If this is not the case, dates for future process review to ensure maintenance of improvement may be set.

The benchmarking process

In addition to reviewing the operational results of the benchmarking study, the progress of the benchmarking process needs to be examined. It is important to capture learning and to develop the benchmark networking capabilities of the organisation. Some of the issues that should be covered in this examination are as follows:

- The organisational strengths and weaknesses identified during the study.
- The need (or otherwise) to improve benchmarking training and skills before the next benchmarking study.
- The level of resistance to or motivation for change within the organisation.
- Effectiveness of communication.
- The levels of commitment displayed by those directly involved in the benchmarking exercise and steps that may be taken to overcome any resistance in future.
- The level of understanding and acceptance of benchmarking practice within the wider organisation.
- Any changes or adjustments that may be made to the benchmarking process to make it more effective in future.

9.7.2 Maintaining the momentum of benchmarking: re-calibration

Re-calibration refers to the formal review of the benchmark process to ensure its continued validity. This may be carried out as part of the annual business planning cycle, through specially targeted assessment studies or by examination of routinely collected information. As industry practices evolve, re-calibration is necessary to ensure that the organisation keeps up with changing conditions in addition to improving the maturity of its benchmarking philosophy. Companies in less dynamic industries may opt not to re-calibrate on an annual basis but choose a longer timeframe that is better suited for both the industries and their operations.

The starting point for re-calibration is often an internal study that would identify the nature of the gaps in information. The study may also determine changes in attitude to benchmarking since the previous review. The review should also take into account approaches and results from other benchmarking exercises within the organisation as this may identify new and more efficient practices.

The style of the review will depend largely on the maturity and culture within the company and could range from formal questionnaires and audit to less formal information-gathering techniques.

When the internal review has been completed, the benchmarking partner and/ or another best practice organisation may be visited as part of a broader review. The process to be adopted in the external review should be more or less the same as the benchmarking process described in this chapter.

Dedication and constant review of the benchmarking initiative should ultimately lead to the institutionalisation of benchmarking within the organisation. At this point, an overwhelming majority of staff would routinely seek best practices and ensure continuous improvement. It is likely that less of a push will be required from management when employees are able to take such initiatives.

However, management will continue to play an important supporting role in ensuring that benchmarking continues to deliver benefits to the organisation. They will need to recognise and reward employees for their efforts. They also need to ensure that adequate resources are made available to support the ongoing efforts in respect of benchmarking. Some of the ways in which the organisation can manage its benchmarking initiative continuously are described in the next section. It is worth noting that benchmarking can only continue to be successful for as long as management regards it as important to the overall business strategy.

9.8 Managing the benchmarking process

Benchmarking is a change tool that can be aligned to other initiatives within the company (e.g. business process engineering, problem solving). Irrespective of how it is employed, benchmarking needs to be systematically managed to ensure its effectiveness. While the company will get better at benchmarking over time, there will also be a cost and resource implication for the company. The ability of the company to improve continuously by using benchmarking techniques will depend largely on how the initiative is systematically managed and encouraged to become a component of everyday working. While there is no single way to manage a benchmarking initiative, some of the findings from organisations that are experienced at benchmarking are presented here.

9.8.1 Leadership of a benchmarking programme

Without the support, commitment and involvement of senior management, not only might there not be enough resources for undertaking benchmarking activities but recommendations from benchmarking studies may not be implemented. The supporting role that management can play in institutionalising benchmarking can be split as follows:

- *Initiating benchmarking*. Benchmarking should be seen as a tool to assist the attainment of the organisation's mission and should be linked to the corporate strategy. In order to apply this to best effect, it is also advisable to determine expectations and set measurable objectives from the outset. This will send a clear message through the organisation that benchmarking is not only valued by the leaders of the organisation but will play a vital part in the organisation's drive to achieve excellence. If the company intends to use benchmarking continuously, it may be worthwhile to determine protocols for managers and benchmarking teams. This will promote uniformity in the use of benchmarking and ultimately make transferability of personnel and learning experiences seamless.
- *Supporting benchmarking*. Benchmarking teams should be knowledgeable and influential enough to drive change. The members of the teams need to be carefully selected to reflect not only a natural understanding of the relevant processes, but also to have teamwork skills that complement each other and create balance within the group. There may be a requirement to provide training and information before the ideal team dynamics are achieved. In organisations where many benchmarking studies may take place simultaneously or over a relatively short period of time, a benchmarking 'czar' may be appointed to oversee the overall initiative. This person would typically be a very senior manager, who will also be responsible for providing support directly to the benchmarking teams. At a process level, the organisation needs to define ways of determining the right issues for benchmarking. Processes for determining ideal benchmarking partners for a range of issues should be specified and the prospective partners stored on a database.
- *Maintaining benchmarking*. Although it can be argued that once benchmarking is accepted as the norm, it becomes self-sustaining, complacency and inability to innovate may gradually erode the value that

the organisation gets from benchmarking. The organisation should routinely study benchmarking programmes in other companies in order to identify possible beneficial refinements to the process. It is also important to celebrate successes as a way to maintain awareness and provide increased impetus. There should also be reward and recognition for successful benchmarking teams. It should, however, be noted that not all benchmarking exercises will be successful and such circumstances should be used as an opportunity to commend and reassure teams rather than punish them. Finally, management must continue to walk the talk and as much as possible, remaining personally involved in benchmarking exercises.

9.8.2 Stages of benchmarking development

As indicated earlier, an organisation is likely to get better the more it undertakes benchmarking exercises. There are at least three developmental stages to benchmarking maturity. It is important that organisations are aware of the three stages so that they can assess where they are in benchmarking and where they need to go next.

Stage 1 – Evaluating the relevance of benchmarking

This is primarily an exploratory stage for the organisation. It is unlikely to have carried out a benchmarking exercise before and there may be some difficulty in understanding the relevant methods and principles. The organisation is not likely to understand fully the need for benchmarking and how it relates to overall improvement. This may be compounded by poor communication through the organisation. There may also be a lack of conviction and commitment to benchmarking right from the shopfloor through to management. Other characteristics of this stage are:

- Insecurity about the consequences that benchmarking will have on the organisation.
- Reluctance to share information and consequently lose 'control'.
- Unwillingness to participate in benchmarking activities because it 'takes us away from our job'.

Stage 2 – Entrenching benchmarking within the organisation

This is primarily a transition stage. It is perhaps the most important of the three stages as the long-term survival of the benchmarking will depend on the successful negotiation of this stage. Initially, this stage will be characterised by a lack of uniformity and a lack of focus across the organisation. Some benchmarking exercises will have been carried out, but there are likely to be different levels of understanding, awareness and commitment to benchmarking. It is also likely that many people within the organisation will not have the training and skills required to participate in benchmarking. There may also be personal, professional, operational and interest group conflict.

As management support (see previous section) takes ground, most of these problems will be overcome. The increasing focus on external partners (especially from the same industry) may become the new source of resistance. However, there is an increased understanding of the purpose of benchmarking and this may result in better cohesiveness within the organisation.

Stage 3 – Maturity in benchmarking
This is the final stage and it indicates that the organisation has achieved maturity in benchmarking. The organisation will, typically, have carried out many benchmarking exercises and will have developed an effective approach to maximising the benefits of these exercises. The new challenge will be to avoid complacency and to maintain enthusiasm for the programme. It is also feasible that, at this stage, less developed or experienced organisations will be looking to benchmark and may require some measure of support.

9.8.3 Future trends in benchmarking

Over the past ten years, there has been an increasing number of companies undertaking benchmarking studies. This trend is likely to continue as competition, globalisation and the need to satisfy stakeholders make increasing demands on the performance of organisations. The importance and benefits of opening up and sharing information with best practice organisations is taking root, and the likelihood is that this trend will continue. Increasing awareness and subscription to the various business excellence and quality award models will also play a significant part in the sharing and dissemination of best practice.

However, research at Leatherhead Food RA has indicated that the UK food industry lags behind other industries in the awareness and use of both benchmarking and business excellence models. This may be attributed to the following factors:

- The food industry is not as globalised as other industries (such as aerospace and automobile industries) and as such is less exposed to differing business practices.
- Many food sector organisations in the UK are small to medium-sized enterprises, which may struggle to provide resources for benchmarking (especially in a low-margin industry).
- Benchmarking exercises are medium- to long-term programmes and are unlikely to appeal to organisations that require short-term bottom line results.

Encouragingly, the indications are not all negative. Many food companies in the UK participate in sharing initiatives such as those promoted by the DTI, Cranfield University (Management Today) and the Best Practice Club. In particular, Leatherhead Food RA's business excellence and benchmarking activities continue to attract commendable interest from the industry. To date, about 250 organisations have joined the business excellence self-assessment programme while the Food and Drinks Industry Benchmarking Club continues to move from strength to strength.

In conclusion, the increasing demand from consumers will drive the need for innovation and flexibility from the food industry. This in turn will promote the seeking of best practice from within and outside the food industry and consequently promote the awareness and adoption of benchmarking.

9.9 Sources of further information and advice

Listed below are organisations that may provide assistance with respect to obtaining benchmarking information or data.

The Department of Trade and Industry
Kingsgate House, Bay 511
66–74 Victoria Street
London SW1E 6SW

www.dti.gov.uk
www.fitforthefuture.org.uk

The DTI has provided financial support for a range of benchmarking initiatives including the production of guides and other publications.

Management Today (Cranfield School of Management)
Cranfield University
Cranfield, Bedford
MK43 0AL

Tel: +44(0)1234 721122
Fax: +44(0)1234 751806

Management Today, in association with Cranfield School of Management, runs an annual award scheme across all industries for Britain's best factories. All participating companies receive a free confidential and detailed report, which compares their performance with industry standards.

Inside UK Enterprise
Status Meetings Limited
Festival Hall
Petersfield
Hampshire GU31 4JW

www.iuke.co.uk
Tel: +44(0)1730 235015

Fax: +44(0)1730 268865

Inside UK Enterprise is a DTI scheme that provides an opportunity for senior managers to visit over 120 leading companies to gain a better understanding of the processes, technology and strategic issues that have helped build successful businesses.

British Quality Foundation
32–34 Great Peter Street
London SW1P 2QX

www.quality-foundation.co.uk
Tel: +44(0)20 7654 5000
Fax: +44(0)20 7654 5001

The BQF runs the annual British Quality Award. It is possible to obtain from them past award winners' submission documents.

The Industrial Society
Robert Hyde House
48 Bryanston Square
London W1H 7LN

www.indsoc.co.uk
Tel: +44(0)20 7479 2127

The Society publishes *Managing Best Practice*. This monthly report focuses on a different topic each month with the aim of being an authoritative and practical benchmark.

Institute of Quality Assurance
12 Grosvenor Crescent
London SW1X 7EE

www.iqa.org
Tel: +44(0)20 7245 6722
Fax: +44(0)20 7245 6755

The IQA is a professional body that promotes quality practices. Membership of the organisation enables access to the National Quality Information Centre.

The Food and Drinks Industry Benchmarking Club
Leatherhead Food Research Association
Randalls Road
Leatherhead
Surrey KT22 7RY

www.lfra.co.uk
Tel: +44(0)1372 376761
Fax: +44(0)1372 386228

The Club is a forum for leading food companies to benchmark their performance and share best practice. It aims to be the flagship for business excellence within the food industry.

The Benchmarking Centre Ltd
Truscon House
Station Road
Gerrards Cross
Bucks SL9 8ES

www.benchmarking.co.uk

The Centre is involved in a number of activities aimed at promoting best practice, helping to identify benchmarking partners and facilitating the exchange of information.

The American Productivity & Quality Center
123 North Post Oak Lane
3rd Floor
Houston
TX 77024-7797
USA

www.apqc.org

The Center has created the International Benchmarking Clearinghouse to provide a resource for organisations interested in using benchmarking as a tool for breakthrough improvement.

IFS International Ltd
Wolseley Business Park
Kempston
Bedford MK42 7PW

Tel: +44(0)1234 853605
Fax: +44(0)1234 854499

IFS also manages a Benchmarking Clearing House in addition to producing publications on a range of topics, including best practices and performance information.

Best Practice Club
Wolseley Business Park
Kempston
Bedford MK42 7PW

www.bpclub.com
Tel: +44(0)1234 853605
Fax: +44(0)1234 854499

The Club's activities include organising best practice visits and facilitating networking between member companies. The Club publishes a monthly magazine that usually includes case studies and information for success.

The PROBE Initiative
Manufacturing Division
Confederation of British Industry
Centre Point
103 New Oxford Street
London WC1A 1DU

www.cbi.org.uk

The initiative is a benchmarking/self-assessment programme that benchmarks manufacturing performance against that of other manufacturers.

European Foundation for Quality Management
Avenue des Pleidas 15
Brussels
B-1200 Belgium
www.efqm.org

The EFQM manages the European Quality Award and promotes quality management in Europe.

9.10 Further reading

The following publications have been referred to in the writing of this chapter and are recommended for further reading.

ADEBANJO, O. and KEHOE, D. (June 1999), LFRA Benchmarking Training Seminar. Presented to UK Food Industry Benchmarking Club.

CAMP, R.C. (1989), *Benchmarking – The search for industry best practices that lead to superior performance.* ASQC Press, Wisconsin.

CAMP, R.C. (1995), *Business Process Benchmarking – Finding and implementing best practices.* ASQC Press, Wisconsin.

CODLING, S. (1992), *Best Practice Benchmarking.* Gower Publishing, Hampshire, UK.

COOK, S. (1995), *Practical Benchmarking – A manager's guide to creating a competitive advantage.* Kogan Page, London.

MANN, R., ADEBANJO, O., and KEHOE, D. (1999), 'Best practices in the food and drinks industry', *British Food Journal*, Vol. 101, No. 3, pp. 238–53.

WATSON, G.H. (1993), *Strategic Benchmarking – How to rate your company's performance against the world's best.* Wiley, New York.

10

Environmental audits and life cycle assessment

B. Mattsson and P. Olsson, The Swedish Institute for Food and Biotechnology, Gothenburg

10.1 Introduction

Today, environmentally friendly production within agriculture and the food industry is widely recognised. Agriculture, food processing, transport, etc., all contribute to the total environmental impact. The environmental manager in the food industry must take this into account, i.e. look not only at the in-factory environment but also at the whole food chain.

Sustainable production and consumption of food requires the incorporation of environmental issues alongside more traditional points of interest such as cost, performance, service, quality, profitability, and customers' preferences. Decision making processes take one or more of these issues into account, but the priority given to each of these concerns will differ from one stakeholder to another. Consumers have considerable freedom in decision making; they can choose from an extremely broad range of food products. On the other hand, agriculture is much more limited once a decision has been made to produce a certain type of product. Compared with consumers, agriculture is much more influenced by external factors (climatic conditions, soil type, support of certain activities by means of subsidies from authorities, requirements put forward by the food industry and retail companies, etc.) (Ceuterick *et al.*, 1999).

Environmental awareness is rather new – it has emerged mostly during the last decade. The reasons for establishing and developing an environmental programme are many, ranging from the welfare of future generations to explicit demands from customers. There are also numerous examples of environmental programmes and financial benefits that are closely linked. Lower energy use and water consumption, for example, are often reflected in a marked reduction in costs (see Fig. 10.1). Authorities can have a significant impact on the overall

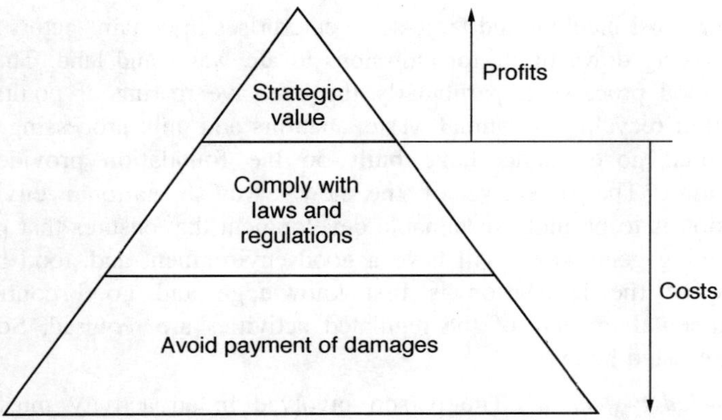

Fig. 10.1 Good environmental management increases the strategic value of a company and unnecessary costs can be avoided.

environmental performance of food chains via regulations and monitoring. At the company level, environmental awareness can be raised through the integration of environmental considerations in quality management and self-regulation and control.

10.2 Environmental legislation

Member states of the European Union are required to integrate the provisions of any directives passed in the European Parliament into national law within two years. Where a country fails to integrate any provision, the Commission takes action in the European Court of Justice (C&C, 2000). The directorate for environmental issues is DGXI: Environment, Nuclear and Civil Protection, while the European Environment Agency (EEA) provides information to policy making agents and the public. The information it provides aims to assist the European Union to improve the environment and move towards sustainability.

European environmental legislation started in the 1970s with directives on waste, conservation and biodegradability of surfactants. The directives that are currently in force are split into sections that include water protection and management, the monitoring of atmospheric pollution, chemicals, industrial risk and biotechnology, waste management and clean technology, and international co-operation (C&C, 2000). The EU instituted its first environmental action programme in 1973. The 5th Environmental Action Programme (1993–2000) was entitled 'Towards sustainability' and sought to foster a number of environmental best practice programmes. Specific environmental directives include Directive 97/11/EC covering environmental impact assessment (EIA) which requires a formal EIA in certain types of food processing including water management on farms, intensive pig or poultry rearing, and fish farming. Council Directive 96/61/EC established a system for integrated pollution prevention and control (IPPC) requiring most medium and large-scale enterprises in certain sectors to obtain a permit laying down limits for emissions to air, water and land. The Directive covers food processing, particularly the intensive rearing of poultry or pigs, disposal or recycling of animal waste, abattoirs and milk processing.

National governments have built on the foundation provided by EU regulations. The objective of the new Swedish national environmental legislation is to promote sustainable development that ensures that present and forthcoming generations will have a good environment and good health. The essence of the legislation is that knowledge and good control of the environmental impacts of the regulated activities are required. Some of the rules are listed below:

- *Knowledge demand.* The person involved in an activity must have the necessary knowledge of how the environment and human health are affected and can be protected.
- *Burden of proof.* The person involved is the one who has to prove that the rules are followed.
- *Precautionary principle.* There is an obligation to take action if there is a risk of damage.
- *Best available technology.* The best available technology should be used to avoid damages and inconvenience.
- *Sustainability principle.* Natural resources and energy should be used as efficiently as possible.
- *Substitution principle.* If possible, a chemical or a biotechnic organism should be replaced by a less hazardous one.

This means that a company is obliged to have the necessary knowledge of the environmental impact of its activities and it must actively seek production methods or solutions that are less hazardous to the environment and to human health.

10.3 Environmental management systems (EMS)

Company self-regulation is covered by a number of initiatives such as 'Responsible Entrepreneurship' (International Chamber of Commerce), 'Responsible Care' (focusing on the chemical industry) and 'Product Stewardship'. In each of these initiatives, care for the environment is at stake. As well as self-regulation, efforts have to be made to integrate environmental considerations in quality management systems such as ISO 9000, ISO 14000 and HACCP. ISO 9000 covers product quality in general, whereas ISO 14000 focuses on environmental management in the company.

ISO 14000 is a series of standards helping companies and organisations all over the world to establish and maintain structured, harmonised and systematic environmental management systems. The series of standards consists of two parts: one aimed at how to organise systems and the other focusing on products.

There are three organisation oriented standards:

- environmental management systems
- environmental audit
- environmental performance.

At present there are also two product oriented standards:

- environmental labelling and declarations
- life cycle assessment.

There are two documents governing independent third party certification:

- environmental certification of companies and organisations (Environmental Management Systems according to ISO 14001:1996)
- environmental certification of products and services (type III Environmental Product Declaration according to ISO 14020:1998).

ISO 14001 is the most important in the series of standards. The Environmental Management System it describes provides the basis for setting up an environmental programme, an organisational structure with shared responsibility among the personnel and increased competence. The system includes procedures for document and activity regulation.

ISO 14001 offers a common, harmonised approach for use by all organisations, wherever they are in the world. Designing processes/equipment to include environmental considerations requires an evaluation of all aspects of a product or service (ideally, from 'cradle to grave', although this is not explicitly stated by ISO 14001). It is only through the establishment of an Environmental Management System (EMS) that an organisation can, over time, monitor and control these aspects. In other words, an effective programme of design for the environment requires an EMS.

There are now many examples available (and frequently reported in the environmental press) which show that there are considerable opportunities to reduce environmental impacts through innovations in product design, processes and methods of operation. With increasing public awareness of environmental issues, it is becoming more likely that environmental credentials will play a part in customer loyalty. For example, environmental aspects are now commonly being incorporated in labelling and packaging of many mainstream products.

10.4 Auditing an EMS

ISO 14001 defines an environmental audit as:

> A systematic, documented verification process for objectively obtaining and evaluating evidence to determine whether specified environmental activities, events, conditions, management systems or information about these matters conforms with audit criteria, and communicating the results of this process to the client.

Environmental audits are intended to quantify environmental performance and define what needs to be done to sustain or improve it. Environmental audits are typically divided into three types:

1. A liability audit encompasses compliance with an organisation's legal obligations. Liability audits may result from a pre-acquisition investigation or one linked to funding requirements (for example investment through an ethical fund).
2. A management audit focuses on verifying that an EMS is meeting its stated objectives. It may be conducted by an organisation to measure its own performance as part of its EMS, or by a third party providing an objective assessment of the organisation's strengths and areas for improvement
3. An activity/issues audit concentrates on a detailed investigation of a chosen area, e.g. energy use.

ISO 14001 provides guidance on the general conduct of audits. Whether conducted by an internal management team to test the robustness of a company EMS, or by a third party, an audit should begin with clear and agreed objectives:

- if the audit is restricted to legal compliance, the terms of reference will be defined by current regulations and standards
- a management audit will be guided by the stated objectives of the organisations EMS. This may involve identifying best practice in the industry as a whole if company performance is to be judged against its competitors. Objectives can then be agreed which test performance against best practice benchmarks

10.4.1 Preparing for an audit

It is essential to agree the scope of the audit at the outset. This might be the entire company, a particular function (an activity audit), or a particular site. The audit will also need agreed priorities, for example, focussing on particularly important or environmentally-sensitive sites for initial inspection. Finally, there will need to be a staged programme for the audit, identifying the resources required, including the information and assistance required from relevant company staff, and the sequence of audit activities from information gathering to site visits and the completion of the final audit report.

Once a clear framework for the audit is agreed, preparation work can begin. This will involve:

- identifying and selecting the people required for the audit team
- gathering existing information

Information required includes:

- the company environmental policy
- management structures and key environmental responsibilities
- the documented EMS and its supporting procedures and records, including

emergency procedures to deal with incidents such as unauthorised discharges, on-site spills, fires or other emergencies
- background information on sites, processes and products, including, where relevant, transport and distribution
- training records and programmes supporting the EMS
- records of significant non-compliances, e.g. unauthorised discharges
- previous audit reports.

An assessment of this information will suggest key areas for particular scrutiny and will inform a pre-audit or pre-survey questionnaire (PSQ). This will cover outstanding issues and form the basis for a site survey. As an example, if not covered in existing documentation, a PSQ may cover the following aspects of a particular site:

- topography of the site and adjacent area
- geology
- hydrology
- potential pathways for pollution
- land use adjacent to the site
- significant contamination risks, for example to people or local wildlife habitats.

It may then cover more specific aspects of company activity, for example waste discharge, air emissions, the handling of hazardous materials, solid waste management and spill controls. Typical questions covering effluent and waste handling are outlined in Table 10.1.

10.4.2 The site visit

A site visit is the last stage in conducting an audit investigation and should focus on issues arising from the PSQ and other information gathered. It is essential that the schedule and sequence of activities should be agreed in advance so that key personnel and data are available and that, where necessary, protective clothing and special access are arranged. A site visit provides an opportunity to see appropriate procedures controls in action, to question staff directly on their knowledge and understanding of their responsibilities within the EMS, to investigate issues identified in the PSQ in more depth, and to identify potential environmental problems.

The treatment of waste water can be taken as an example of issues covered in a site visit. If not already provided, the auditor might request to see such evidence as:

- waste disposal discharge consents
- water supply and treatment records
- effluent monitoring procedures and results
- records of non-compliance and corrective action taken.

Table 10.1 Audit questions on effluents and waste

- What effluents and wastes are generated on site?
  - solids
  - effluents (to sewer or water)
  - emissions (to air)
- Where are they generated?
- What quantities, qualities and properties of these wastes are generated?
- What frequency and with what variations?
- Where do the wastes go?
- What treatment or recycling technologies are in place?
- Who is in charge of managing these wastes and effluents?
- Are any contractors used?
- What physical controls are in-place on site?
- What in-house monitoring takes place?
- Which personnel and what monitoring equipment is used?
- At what frequency and to what extent does monitoring occur?
- What records are kept?
- What procedures are in place for dealing with non-compliances?
- Is there an emergency plan to cover accidents?
- What is the maintenance schedule for:
  - process plant
  - effluent and waste disposal, storage or recycling mechanisms
  - monitoring systems
  - pollution abatement technology
- What legal controls apply?
- Who are the control authorities?
- How frequently do they visit?
- What procedure is there for contacting them in an emergency?
- What licences, consents, permits and authorisations exist?
- What is the site's compliance record?
- Has the site suffered any:
  - complaints
  - warnings
  - legal actions
- Have any complaints been acted on and any necessary remedies put in place?
- What environmental insurance exists and what does it cover?

The auditor might then complete a site assessment of all discharge points and drains. Typical problem areas to look out for might include: process water in the workplace making an unauthorised discharge to drain, poor maintenance of drains or storage facilities, failure to follow proper sampling procedures, inoperative or poorly-maintained monitoring equipment, inadequately trained staff and failure to take corrective action to deal with a non-compliance. Finally, the auditor might conduct a series of interviews with key management and operational staff, using open-ended questions to test their understanding of their EMS responsibilities.

10.4.3 The audit report

The final stage in the audit is the preparation of the audit report. It is advisable to check the draft report beforehand with the personnel directly involved in case changes have already been made or points need clarification or correction. The report should be a clear and concise statement, setting out:

- the original objectives and scope
- data used
- assumptions made and techniques used
- a non-technical summary
- quantitative and technical data summarised, wherever possible, in tabular or diagrammatic form
- conclusions
- recommendations for action.

An audit report should be accompanied by a meeting to review the findings, clarify issues, agree a draft timetable for improvements and, where required follow-up meetings to check compliance with recommendations.

10.5 Other environmental assessment methods

There are a number of different environmental tools that may be used to obtain environmental improvement in the food chain. Some tools are of the utmost importance for specific problems in the chain, e.g. risk assessment may be used for assessing the risk to humans and the environment from the application of pesticides in agriculture. Other tools may analyse the whole chain, e.g. LCA (life cycle assessment) and MFA/SFA (material flow analysis/substance flow analysis). A comprehensive guide to analytical tools for environmental design and management is available (Chainet Guidebook, 1999). Some of the tools are described below.

- *Checklists* are qualitative tools that help with environmental management, design, setting eco-labelling criteria, etc. They cover various aspects such as recycling possibilities, minimising harmful substances, and so on. Checklists used for specific purposes, such as design, may be customised for a specific company or sector.
- *Cost–benefit analysis* is an economic tool that can support decisions on larger investments from a social, as opposed to a firm's, point of view. It has been developed as a tool to address the shortcomings of a purely market oriented analysis of costs and benefits. Contrary to other tools for environmental decision support, cost–benefit analysis can take into account the time horizon of effects.
- *Environmental risk assessment (ERA)* offers a comprehensive evaluation of the potential environmental impact, or rather the probability that damage or adverse effects will occur.
- *Material flow analysis/substance flow analysis (MFA/SFA)* monitors systematically the physical flow (in terms of mass units) of materials (e.g.

substances, raw materials, products, wastes, emissions to air, water or soil) through extraction, production, use and recycling, and final disposal in a specific region.

- *Material intensity analysis (MIA)* is used to quantify the whole life cycle requirements of primary materials for products and services. Sometimes the method is used as a screening step for an LCA study.
- *Life cycle assessment (LCA)* is a method for analysing and assessing the environmental impact of potential substitution of material goods and services, taking into account their entire life cycle. LCA is one of the most important developments in environmental assessment since it gives businesses the opportunities to anticipate environmental issues and design environmental factors into new products and processes, rather than just manage the environmental impact of their existing operations.

There may be several reasons for carrying out an LCA, ranging from improvement of environmental performance to product development and development of new legislation. The demands for the LCA, both with respect to the methodology and to the data, may vary accordingly. LCA is but one tool to generate environmental information. It allows the integrating and aggregating of bits and pieces of environmental information into a more limited number of environmental scores. Developing sustainable food production and consumption systems requires the combination of information coming from LCA with that resulting from other assessment tools. LCA has to be combined with other assessment tools to address those issues outside its scope, such as ethics, social considerations, customer preferences, economic feasibility, risk assessment and regulatory issues. The rest of this chapter looks at the key aspects of LCA methodology and how it can be used to audit processes and products within the processing.

10.6 Introduction to LCA

The first European LCA studies of food products were performed at the beginning of the 1990s. Universities, institutes and companies have participated in the development of LCA methodology and the application of LCA to different types of products. SETAC, the Society of Toxicology and Chemistry, has played an important role in the development of LCA methodology and now LCA has become a part of the environmental management system ISO 14000. The ISO certification is very important for the credibility of LCA and for the future use of LCA in companies.

LCA has primarily been used for product development. In the food area studies have often been restricted to the packaging system but studies of complete product systems are becoming more common. The environmental information gained in an LCA is of course only part of the information needed by the decision makers in product development. However, it is increasingly important to be able to show an awareness of environmental issues towards consumers, investors, etc. There is also a clear potential for cost savings when the use of energy, water and other resources can be reduced.

10.6.1 LCA working procedure

LCA is a method used to investigate and assess the environmental impact of a material, product or service throughout its entire life cycle from raw material acquisition through production, use and disposal (see Fig. 10.2). The product system studied is delimited from the surrounding environment by a system boundary. The energy and material flows crossing the boundary are accounted for as:

1. resources used for production, transportation, etc. (inputs), and
2. emissions and waste leaving the product system and entering the surrounding environment (outputs).

The parameters are often numerous, which can make interpretation difficult. Hence, emissions contributing to the same environmental impact (impact category) are aggregated to facilitate interpretation of the results.

The procedure for making an LCA consists of four phases (Fig. 10.3). In the first phase, the *goal and scope definition*, the purpose of the study and its range are defined. In the goal and scope definition important decisions are made concerning boundary setting and definition of the functional unit (i.e. the reference unit). During the *inventory analysis*, information about the product system is gathered and relevant inputs and outputs are quantified. In the *impact assessment*, the data and information from the inventory analysis are linked with specific environmental impacts so the significance of these potential impacts can be evaluated. Finally, in the *interpretation* phase, the findings of the inventory analysis and the impact assessment are combined and interpreted to meet the previously defined goals of the study. Formalised, quantitative weighting methods are available for the aggregation of either inputs and outputs or environmental effects into one index. These methods originate from the social sciences, since the values concerned are held by people within the social system. A review of weighting in LCA has been published by Bengtsson (2000).

10.6.2 System boundaries

The choice of system boundaries has been discussed by Tillman *et al.* (1994). Ideally, the system boundary should be between the technological system and nature. In agricultural production it is difficult to make this delimitation since production takes place in nature. For example, it has been discussed whether the soil should be regarded as part of the technological system or not. Delimitation in time and of the geographical area must also be made, and boundaries established between manufacturing of the product studied and the production of capital goods. In LCA calculations, the industrial production of capital goods, such as machinery and buildings, is normally left out. The reason is that the capital goods often have a long lifetime, which means that the environmental

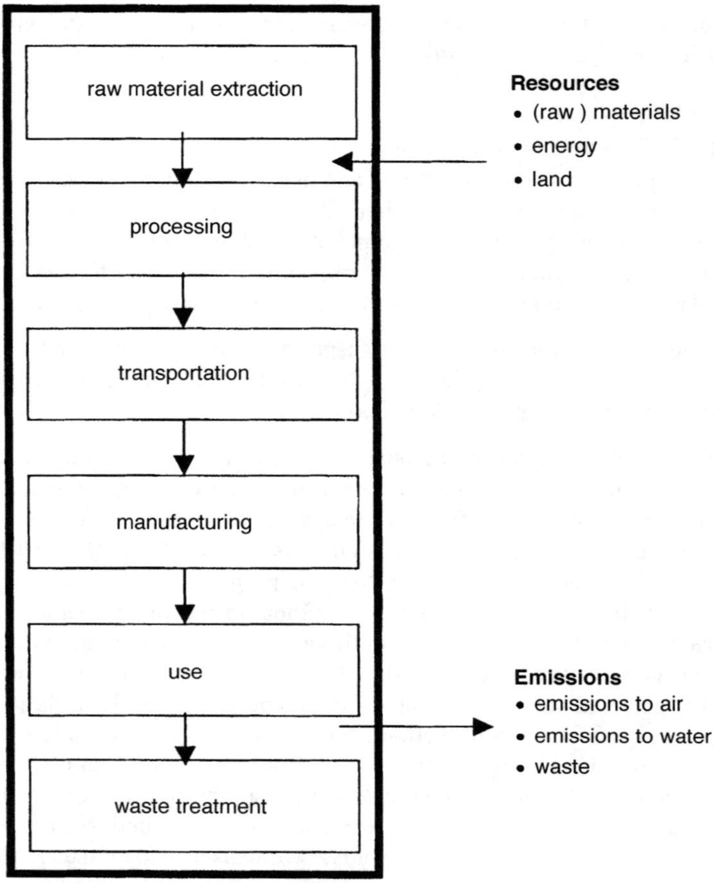

Fig. 10.2 Principal life cycle steps included in an LCA. The product system studied is delimited from the surrounding environment by a system boundary and the energy and material flows crossing the boundary are accounted for.

burdens of their production would be divided among a large quantity of products: for just one product, the result is likely to be negligible.

10.6.3 The functional unit

The definition of the functional unit is determined by the specified main function of the product system under study. The functional unit should be a relevant, well defined and strict measure of the service that the system delivers; it is the basis for the analysis. All data is related to the functional unit (Lindfors *et al.,* 1995).

In most LCA studies of food products, the mass of a specific product has been defined as the functional unit, e.g. 1 kg of bread from a bakery or 1 kg of apples

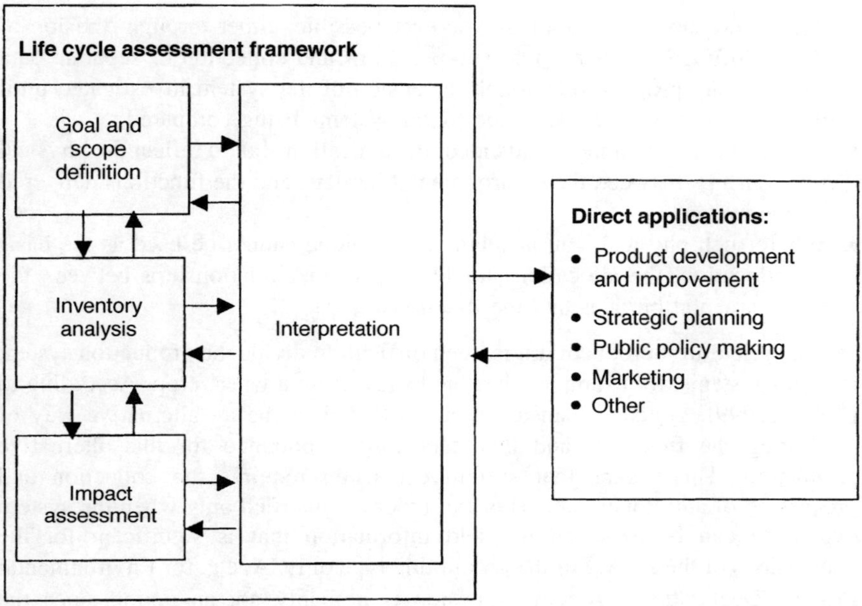

Fig. 10.3 Procedure and application of LCA studies. (Source: ISO 14040: 1998.)

from the greengrocers. As pointed out by Andersson (1998), there are various parameters relevant to the function of a food product: the content of various nutrients and fibres, the calorific value, shelf-life, taste, smell and appearance. A minimum requirement must be that a given food is hygienically and toxicologically safe. The definition of the functional unit must be determined by the goal of the study. Andersson points out also that, when various food products are to be compared, it seems relevant to take into consideration their role or function in the diet, for example the content of proteins for meat or fish.

10.6.4 Allocation

Allocation of environmental burdens may be necessary, for example when the same process yields more than one product, i.e. functions. There are plenty of examples of this type of multifunctional process in agriculture and food processing, for instance dairy cow production yields both milk and meat, vegetable oil crops yield both oil and feed and, when cheese is produced, the process also yields whey.

Allocation procedure in life cycle assessment is a separate field of research that has been addressed by Ekvall (1999). Ekvall gives a thorough overview of allocation methods, of which only a few are addressed here. According to the international standard the following approaches should be used in the following order of preference (ISO 14041:1998).

- Allocation should be avoided, wherever possible, either through division of the multifunction process into sub-processes and collection of separate data for each sub-process, or through expansion of the system investigated until the same functions are delivered by all systems being compared.
- Where allocation cannot be avoided, the allocation should reflect the physical relationships between the environmental burdens and the functions delivered by the system.
- Where such physical, causal relationships alone cannot be used as the basis for allocation, the allocation should reflect other relationships between the environmental burdens and the functions.

For agricultural production it is often difficult to divide the production system into sub-systems, for example wheat and straw from a wheat crop. According to Ekvall (1999), system expansion requires that there be an alternative way of generating the functions and that data can be obtained for this alternative production. This means that system expansion requires the collection and processing of additional data. This extra work is justified only when the system expansion can be expected to yield information that is significant for the conclusions of the LCA. For the accounting type of LCA, e.g. for Environmental Product Declarations, system expansion is probably not an option since the environmental burdens of other products would be involved.

10.6.5 Data collection and data quality

When interpreting the results of an LCA, it is important to have an understanding of the quality and uncertainty of the inventory data. A life cycle study is a summary of a large amount of input data of varying quality. Usually, the uncertainties cannot be quantified; however, they can be expected to be large. The cumulative effects of uncertainties in both inventory and impact assessment are potentially very significant for the overall results. According to Barnthouse *et al.* (1997), there is a clear need to evaluate the influence these uncertainties have upon final (and intermediate) LCA results. Although this is difficult to achieve, an effort should at least be made to present information about how representative the input data are, for example, and how the uncertainties may have influenced the results. For the case studies, Barnthouse made comparisons with other data sources whenever possible; however, this could not be done for specific production data provided by industry.

10.6.6 The use of LCA

According to the ISO standard, LCA can be useful for various purposes:

- identifying opportunities to improve the environmental aspects of products
- decision making in industrial companies or other organisations, such as strategic planning, setting priorities, or product and process design or re-design

- selecting relevant indicators of environmental performance, and
- marketing, such as environmental claims, eco-labelling schemes, or Environmental Product Declarations.

LCA can be useful in auditing since it provides a complete picture of the environmental impact of a business's operations. It can, therefore, provide a standard against which to audit and impove current environmental performance. It also requires a new set of auditing skills in verifying a particular LCA or the incorporation of LCA principles into an existing EMS.

10.7 The food life cycle

The primary benefit of LCA is that it can give a relatively complete picture of the environmental profile of the product. The whole life cycle of the product is addressed, to avoid a situation in which problems are shifted between different life cycle phases or between different environmental impacts.

10.7.1 Agricultural production

Plant nutrient supply and emissions are decisive in life cycle assessment studies of agricultural products. The plant nutrient management in an agricultural system affects several LCA impact categories regardless of whether the plant nutrients are applied as manure or fertilisers. The most important resources utilised for fertiliser production are fossil fuels and phosphate ore, and the nitrogen and phosphorus emissions are important contributors to eutrophication and acidification. Since manure is a by-product of livestock production, there is also a need for allocation of the environmental burdens between livestock and plant production.

Normally, the energy used in pesticide production is not one of the major contributors to the total energy use. The process emissions caused by pesticide production are very small compared with the amounts of pesticide products that are spread directly into the environment. According to Guinée (1995), toxicity was a problem category (and it still is) because of the lack of data and knowledge. For the characterisation of ecotoxic and human toxic emissions, data such as vapour pressure and degradation rates for soil, water and air are needed for each chemical and for its metabolites. These data are often lacking because knowledge of these aspects is still limited. Guinée also pointed out that due to the lack of knowledge, the data sets are far from complete in current characterisation methods for toxic and human toxic emissions.

It is clear that further attention should be paid to the impact assessment of toxic compounds in LCA. It also seems clear that more information about the environmental effects of pesticides and other toxic compounds would be desirable and that every precaution should be taken before allowing the use of these compounds. It would be desirable to examine and include not only the impact of the pesticides released in the environment but also pesticide residues in food products in LCA.

LCA is mainly a systems analysis of energy and material flows. It is definitely desirable to combine it with other types of environmental impact. For instance, an agricultural product has a direct impact on the long-term fertility of the soil and on biodiversity. Common features for land use characterisation methods suggested by several authors are: (1) area occupancy as resource utilisation, (2) impact on soil quality and productivity, and (3) impact on biodiversity (e.g. Cowell, 1998; Lindeijer *et al.*, 1998). The impact of land use in LCAs of food products should definitely be included. However, a case study of vegetable oil crops (Mattsson *et al.*, 2000) showed that a thorough investigation is very time consuming and that it can sometimes be difficult to find detailed information about imported crops. Therefore there is clearly a need for simplified methods that can be used when there are numerous ingredients in a food product under study. It is also quite clear that there is an urgent need for co-operation between the different research groups in order to co-ordinate progress of the methodology.

10.7.2 LCA of industrial food processing

Energy use and energy related emissions are the most important contributors to the total environmental impact of food processing. It is mainly heating processes but also chilling and freezing that contribute to the energy use. Fossil fuels seem to be the dominating energy carrier for heating processes; however, electricity is also used in pumps, refrigerators, freezers, etc.

Andersson (1998) showed that food processing and packaging were the most energy demanding phases of ketchup production. However, if the storage time in the household is 12 months, the electricity for the refrigerator consumes more energy than that used in food processing and packaging.

In a study of carrot purée used for baby food (Mattsson, 1999) the energy use in food processing accounted for approximately half of the total energy use for the entire life cycle. Although heating and sterilising processes are very energy demanding, there is a potential for energy recovery. In a study of cereal based baby food products, the evaporation of skimmed milk into milk powder and whey into whey powder were found to be very energy demanding processes. These processes accounted for approximately 70 per cent of the total energy use in food processing and as much as 45 per cent of the energy use in the whole life cycle.

The major water emissions from the food industry are oxygen demanding compounds, such as fats, nitrogen and phosphorus. Oxygen demanding compounds and phosphorus are not normally a problem, since sewage treatment can reduce these emissions efficiently. However, if large amounts of sewage are released into a stream, this might cause eutrophication problems locally. The emissions to water from the food industry have turned out to be small compared with the nitrogen and phosphorus emissions from agriculture.

The energy use for production of cleaning agents, calculated in the baby food study, turned out to be negligible (Mattsson, 1998). However, Weidema *et al.* (1995) have pointed out that some detergents may have effects on human or ecosystem health which should warrant closer investigation. In the baby food study (carrot purée) this was done as a red flag classification. One of the cleaning agents was potentially hazardous to human health (P3-super LA), while sodium hypochlorite and phosphonates are potentially hazardous to ecosystems.

It is important to stress that cleaning agents are used to avoid microbiological problems which can be very dangerous to human health. Nevertheless, hazardous chemicals should be replaced if possible. It is clear that impact assessment methods are needed for cleaning agents as well as other toxic compounds.

10.7.3 Food packaging

Modern packaging is a key element in reducing losses of food and improving the health and safety of people throughout the world. From an environmental point of view, packaging should never be optimised in isolation. According to Bürkle (1998), minimising product loss is more important than minimising packaging because the packaging accounts for only 10 to 15 per cent of the total energy necessary for the food supply. Thus, not using packaging or using over-large package sizes can be the worst solution for the environment. However, efforts should be made to reduce packaging weight without jeopardising food safety. Significant weight reductions of packages have been achieved in the past two decades through better raw materials, improved converting techniques and more efficient design. Very often, the less material and energy used, the lower the cost. In many cases, ecology and economy go hand in hand.

The selection of packaging material plays a key role in both the resource aspect and the emissions. Paper has an advantage over most other packaging materials, since the wood fibre raw material is a renewable resource. A major part of the European paper industry is based upon the reuse of paper. In addition, when the fibres are not suited to re-packaging, they retain an excellent energy content, making them attractive for incineration with heat recovery (Jönsson, 1998).

10.7.4 Food transportation

Most foods are transported by road vehicles (88 per cent); the transporting of food products accounts for 16 per cent of the total transport volume in Sweden (tonkm – the distance (km) an amount of goods (tonnes) is transported). Road vehicles produce significant amounts of harmful emissions such as NOx, CO_2 and VOCs (SEPA, 1999). In the context of food LCA, it is often surprising that the environmental impact of transportation does not contribute more to the total environmental impact. The most important transportation stage in a food system may well be the transport of foods from the retailer to the home of the consumer if he or she goes shopping by car.

10.8 Case studies

Life cycle assessment can be used to answer questions that are interesting from an environmental point of view. For instance, it is possible to identify the sub-systems contributing most to the total environmental impact in a product system and it is possible to compare products or processes with the same function.

10.8.1 Organic versus conventional products

When comparing organic and conventional products, it should be remembered that production is on a different scale in the two methods. Some examples of case studies are given below. The organic and conventional production systems of two baby food products were investigated by Mattsson (1999). The major advantage of organic production was the ban on pesticides, while the major disadvantages were the lower crop yields and the difficulties of avoiding plant nutrient emissions from organic fertilisers. The pesticide use was the major drawback of the conventional cultivation systems, although the higher crop yields resulted in lower environmental impact per kg of product, even when the impact per hectare was the same as in organic production. However, a comparative study of organic and conventional farm milk production showed that the conventional production system, with a high input of imported cattle feed, clearly has a larger environmental impact than an organic, more self-supporting production system (Cederberg and Mattsson, 2000).

10.8.2 Scale of production

Andersson (1998) has reported the results of a case study of bread. One of the objectives of the study was to compare the influence of scale of production. Home baking, a local bakery and two industrial bakeries of different sizes were studied. The home baking system showed a relatively high requirement for energy and water; otherwise, the differences between home baking, the local bakery and the small industrial bakery were negligible.

The scale of dairy milk production was studied by Høgaas Eide (2000). Three different Norwegian dairies were compared. The results showed that the environmental impact of the smallest dairy was significantly higher than for the other two dairies. The explanation was that the process equipment in the small dairy was cleaned more often, thus the energy use per kg of milk was higher.

It is often assumed that small enterprises cause less environmental impact than large companies. The two studies quoted above show that no such conclusion can be made. However, it is important to stress that in studies of existing companies not only the scale of production but also other sub-systems differ among the subjects.

10.9 The benefits of LCA

The results from an LCA study can be used internally in a company as a basis for monitoring progress, and they can also be used externally to communicate results to the wider public. Table 10.2 provides an overview of the most significant goals for carrying out a food LCA, together with illustrative examples.

For the industry, LCA's main use is in product development. It has a number of other potential application areas for the industry; however, the method would have to be further developed. For various purposes (particularly in product design) it is desirable to have a simple set of food related eco-indicators and this is an area which still has to be further developed. LCA information would be helpful for various monitoring purposes, for example:

- to demonstrate product improvement
- to benchmark production processes
- to communicate with the public on progress in environmental work.

In the distribution and sales of food products, there is an ongoing trend towards increased concentration: more big stores and fewer retail outlets. Furthermore, new shopping systems have developed and are gradually being introduced, e.g. telemarketing and teleshopping. Factors which have led to this trend are:

- a minimisation of packaging (waste)

Table 10.2 Overview of some goals for carrying out a food LCA

Goal of study	Examples
Learning and/or awareness-raising	
Provision of information	Teaching
Hot spot identification	Research
	Environmental reports
	Eco-labelling
	Product declarations
	Environmental assessment of farms
Operational decisions	
Short-term system optimising	Product improvement
	Production system improvement
Strategic decisions	
Long-term strategic planning	Development of policy, legislation
	Designing new ranges of products
	Setting priorities
	Defining criteria to be met by suppliers
	Choice between different farming methods (e.g. conventional versus organic)

Source: LCAnet Food, 1998.

- a reduction of energy consumption associated with preservation (storage) and distribution
- better preservation characteristics of the products (preventing food losses, longer shelf-life)
- an increased demand for stock flexibility.

As an environmental management tool, LCA can assist distributors in the selection of optimal systems for distribution (e.g. bulk versus packed goods). Another use is to support a choice between alternative logistic routes (e.g. centralised versus decentralised distribution, truck versus rail). LCA is a tool well adapted for hot spot identification in the supplier–centralised warehouses–shops–consumer chain. It can also be used for the development of criteria for selecting suppliers, and last but not least, to communicate environmental performance to consumers. Other factors besides environmental issues have to be included in the final decision making process: costs, logistics, technical feasibility, customer requirements, etc. Most probably, economics (cost minimisation) play the most important role at present.

The most significant LCA application in relation to distribution and consumption at the moment is product policy, often connected with the EU packaging directive and various eco-labelling schemes. Relevant environmental issues associated with distribution and sales are energy (transportation, storage) and organic waste (food losses).

Individual consumers do not have any direct involvement in using and applying LCA. The demands at this level are represented by different national and international consumer organisations. However, indirectly the results of an LCA study are very interesting for the consumer. Therefore, one of the most relevant issues is how the results of an LCA are communicated to consumers (on the food label next to the ingredients list, on a separate eco-label, leaflet, etc.) (Ceuterick *et al.*, 1999)

10.10 Future trends in LCA

One could say that there are contradictory trends in LCA application. Some advocate life cycle stressor effects assessment (LCSEA) which aims to integrate LCA with other environmental assessment techniques (SETAC, 1997). Only the *potential* environmental impact of a product system is normally calculated in LCA; however, the purpose of the LCSEA concept is to take the actual impact of the emissions into account, including the susceptibility of the local area to the substances emitted. For instance, acidifying emissions in an already sensitive area should be considered more serious than the same emissions in a location less sensitive to acidification. To make this concept feasible, detailed inventory data about the location of the emissions released are required as well as knowledge of the cause-and-effect relationships between emissions and environmental effects. Other LCA practitioners advocate simplified LCA methods to make them less time consuming and costly (Christiansen, 1997). This may be achieved when more generic data are available in databases, if relevant cut-off criteria for minor materials can be formulated, etc. Nevertheless, there is always a risk of losing important environmental information when LCA is simplified.

10.11 References

ANDERSSON, K. 1998. *Life Cycle Assessment (LCA) of Food Products and Production Systems.* Chalmers University of Technology, School of Environmental Sciences and Department of Food Science. Ph.D. thesis. Göteborg, Sweden.

BARNTHOUSE, L., FAVA, J, HUMPHRIES, K., HUNT, R., LAIBSON, L., NOESEN, S., OWENS, J., TODD, J., VIGON, B., WEIZ, K. and YOUNG, J. 1997. *Life Cycle Assessment: The State-of-the-Art.* Society of Environmental Toxicology and Chemistry. Pensacola, FL, USA.

BENGTSSON, M. 2000. *Environmental Valuation and Life Cycle Assessment.* Chalmers University of Technology, Department of Environmental Systems Analysis. Licentiate thesis. Göteborg, Sweden.

BÜRKLE, D.H. 1998. 'Optimising packaging: Fitness for purpose, together with ecological and economic aspects, must be part of the equation', in: Klostermann, J.E.M. and Tukker, A. (eds) *Product Innovation and Eco-efficiency: Twenty-three Industry Efforts to Reach the Factor 4.* Kluwer Academic Publishers.

CAMPDEN & CHORLEYWOOD (C&C) FOOD RESEARCH ASSOCIATION GROUP 2000. 'Environment matters for the food industry', Newsletter, May 2000. Chipping Campden, UK.

CEDERBERG, C. and MATTSSON, B. 2000. 'Life cycle assessment of milk production: A comparison of conventional and organic farming', *Journal of Cleaner Production.* 8, 49–60.

CEUTERICK, D., DUTILH, C. and WRIESBERG, N. 1999. *Demand for Environmental Information/LCA Communication.* SIK-document 137, Theme report in LCAnetfood, EU-project 97-3079, March.

CHAINET GUIDEBOOK. 1999. EU-project CT-97-0472, October.

CHRISTIANSEN, K. (ed.). 1997. *Simplifying LCA: Just a cut?* Society of Environmental Toxicology and Chemistry. SETAC-Europe. Brussels, Belgium.

COWELL, S.J. 1998. *Environmental Life Cycle Assessment of Agricultural Systems: Integration into Decision Making.* University of Surrey, Centre for Environmental Strategy. Ph.D. thesis. Guildford, UK.

EKVALL, T. 1999. *System Expansion and Allocation in Life Cycle Assessment with Implications for Waste Paper Management.* Chalmers University of Technology, Department of Technical Environmental Planning. Ph.D. thesis. Göteborg, Sweden.

GUINÉE, J.B. 1995. *Development of Methodology for Environmental Life-Cycle Assessment of Products with Case Study of Margarines.* University of Leiden. Ph.D. thesis. Leiden, The Netherlands.

HØGAAS EIDE, M. 2000. 'Life Cycle Assessment (LCA) of industrial milk production', Int. J. of LCA (in press).

ISO 14001. 1996. *Environmental Management Systems: Specification with Guidance for Use.*

ISO 14020. 1998. *Environmental Labels and Declarations: General Principles.*

ISO 14040. 1998. *Environmental Management: Life Cycle Assessment – General Principles.*

ISO 14041. 1998. *Environmental Management: Life Cycle Assessment – Goal and scope definition and inventory analysis.*

JÖNSSON, G. 1998. 'Paper packaging designed for recycling', in: Klostermann, J.E.M. and Tukker, A. (eds) *Product Innovation and Eco-efficiency: Twenty-three Industry Efforts to Reach the Factor 4.* Kluwer Academic Publishers.

LCANET FOOD. 2000. Final Document, EU-project 97-3079, SIK Document no. 137.

LINDEIJER, E., VAN KAMPEN, M., FRAANJE, P., VAN DOBBEN, H., NABUURS, G.-J., SCHWENBOURG, E., PRINS, D. and DANKERS, N. 1998. *Biodiversity and Life Support Indicators for Land Use Impacts in LCA.* University of Amsterdam, IWAM ER. The Netherlands.

LINDFORS, L.-G, CHRISTIANSEN, K., HOFFMAN, L., VIRTANEN, Y., JUNTILLA, V., HANSEN, O.-J, RØNNING, A., EKVALL, T. and FINNVEDEN, G. 1995. *Nordic Guidelines on Life-Cycle Assessment.* Nordic Council of Ministers. *Nord*: 20.

MATTSSON, B. 1998. *Life Cycle Assessment (LCA) of Carrot Purée: A Case Study of Organic and Integrated Production.* SIK-Report No. 653.

MATTSSON, B. 1999. *Environmental Life Cycle Assessment (LCA) of Agricultural Food Production.* Swedish University of Agricultural Sciences, Department of Agricultural Engineering. Ph.D. thesis. Alnarp, Sweden.

MATTSSON, B., CEDERBERG, C. and BLIX, L. 2000. 'Agricultural land use in Life Cycle Assessment (LCA): Case Studies of three vegetable oil crops', *Journal of Cleaner Production* 8, 283–92.

TILLMAN, A.-M., EKVALL, T., BAUMAN, H. and RYDBERG, T. 1994. 'Choice of system boundaries in Life Cycle Assessment', *Journal of Cleaner Production* 2, 21–9.

SEPA. 1999. *Svensk produktion med miljön i focus.* (*Environmental aspects of Swedish production*). Swedish Environmental Protection Agency. Report 4962. Stockholm.

SETAC. 1997. *Evolution of a Technical Framework for Life-cycle Impact Assessment.* Short course at the SETAC 18th Annual Meeting, 16–20 November. San Francisco.

WEIDEMA, B.P., PEDERSEN, R.L. and DRIVSHOLM, T.S. 1995. *Life Cycle Screening of Two Products: Two Examples and some Methodological Aspects.* Danish Academy of Technical Sciences.

11

Auditing organic food processors

J. R. Parslow and J. Troth, Soil Association Certification Limited, Bristol

11.1 Introduction

Many processors have recognised the massive and sustained growth in the organic food sector. In order to take advantage of this, or remain competitive, or keep major customers, they have had to give consideration to developing organic processed products. Due to market place changes in the last 2–3 years (Taylor Nelson Sofres, 2001) the practical and commercial viability of organic processing is now much more attractive and organic market growth is running at around 40% per annum.

11.2 Defining organic food processing

Organically grown food is food that is grown according to a set of principles and legally defined standards. Organic food processing can be defined as taking ingredients produced to these standards and converting them into a product that is acceptable and desirable to the consumer whilst maintaining the organic 'integrity' of the ingredients, the product and process. The 'integrity' can be defined as the organic product maintaining its 'identity', being free of cross contamination from other foods and chemicals and not having its organic 'vitality' lost by over processing. Put simply, organic food processing is not just the substitution of organically grown ingredients for non-organically grown ones, but the conversion of a set of production principles into a consumer product. Organic food processing must be viewed as part of a whole food production and supply system and not just a separate function that sits between primary production and consumption.

Organic food processing has to meet many conflicting demands which non-organic products do not have to satisfy. Examples would include:

- having to meet high standards of food safety but have minimal packaging and use minimal additives
- trying to adopt a local ingredient sourcing policy when there is currently a UK shortage of organically produced raw materials
- having to satisfy environmental requirements such as efficient energy use but short production runs in non-dedicated factories leads to greater cleaning requirements and greater processing losses.

To give an introduction to the challenges of organic food processing and auditing within the industry some of the main issues relating to organic processing are described below.

11.2.1 Consumer perception and requirements

Organic food processing must not just satisfy a set of guiding principles and standards, but must also give consideration to consumer beliefs and expectations for both the health and environmental benefits of the food and absence of agri-chemicals and food additives. There is also a desire and belief that organic foods are not processed using undesirable practices and processes and that minimal processing and packaging is used. However, the consumer also requires that organic foods are equally as safe as their non-organic counterparts in microbiological terms and shelf stability. They also increasingly require them to meet similar 'convenience' factors to their non-organic counterparts. Additionally, most organic foods are perceived to be more expensive than their non-organic counterparts (Anon, 1999), although certain organic food products, particularly dairy products, are being priced much more closely to similar non-organic products. There is pressure for prices to become lower to make organic foods affordable for a wider range of socio-economic consumers.

11.2.2 Guiding principles and ethics

The International Federation of Organic Agricultural Movements (IFOAM) has produced a set of principle aims of organic production and processing (Anon, 2000). The aims related to processing are listed below:

- To produce food of high nutritional quality in sufficient quantity.
- To consider the wider social and ecological impact of the organic production and processing system.
- To promote the healthy use and proper care of water, water resources and all life therein.
- To use, as far as possible, renewable resources in locally organised agricultural systems.
- To work, as far as possible, with materials and substances that can be reused or recycled, either on the farm or elsewhere.

- To minimise all forms of pollution.
- To process organic products using renewable resources.
- To allow everyone involved in organic production and processing a quality of life which meets their basic needs and allows an adequate return and satisfaction from their work, including a safe working environment.
- To consider the wider social and ecological impact of the farming system.
- To produce fully biodegradable organic products.
- To progress towards an entire organic production, processing and distribution chain, which is both socially just and ecologically responsible.

11.2.3 Regulation

Organic food production is the only system of food production that is legally defined. Because organic food production is a specific system of production, it is necessary to ensure that there is a credible guarantee of authenticity of organic production methods from primary production to consumption. It is often not known or simply overlooked that the organic inspection is actually a legal requirement and that organic standards and certification are not just 'another accreditation or quality assurance scheme'. Organic food processing is controlled in the EU by a regulatory structure which is applied at operator level by means of a certification system. The certification system is discussed in more detail later.

The United Kingdom organic food industry is controlled by EU Regulations 2092/91 (crop products) and 1804/99 (livestock products), the latter now being incorporated into EU 2002/91 and came into force on 24 August 2000. This is enacted in the UK by the Organic Products Regulations 2001 (SI No. 430) and makes it an offence for an unlicensed operator to refer to, or imply, 'organic' production methods for food products of agricultural origin. The regulation is administered in the UK by the United Kingdom Register of Organic Food Standards (UKROFS) which is funded by MAFF. This body sets UK minimum standards and approves Organic Sector Bodies such as Soil Association Certification Ltd to license organic operations. UKROFS also approves organic inspectors independently of the certification bodies they work for.

The regulatory structure is summarised in Fig. 11.1. The legislation requires that any operator who produces, processes, imports, packs, re-packs, labels or re-labels organic food out of sight of the consumer to be licensed by an approved sector body. If the business only sells prepacked organic products (whether wholesale or retail) then it currently does not need a licence. To demonstrate that an organic product offered for retail sale has been produced by a certified operator, the product label must display the code of the certification body responsible for the retail packing operation. This is required throughout the whole of Europe and must be applied regardless of whether the certifier's symbol is used, it is however not required for imported pre-packed products.

The organic standards (Anon, 1999) cover the entire organic food chain from primary production to storage, distribution, importing, processing and retailing.

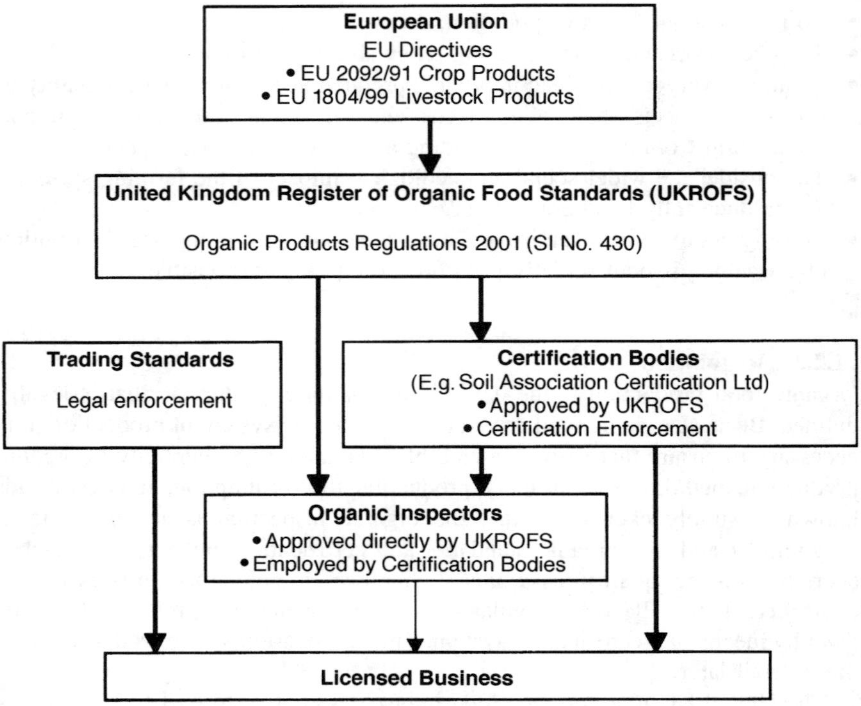

Fig. 11.1 UK organic regulatory structure.

They also cover auxiliary agricultural industries such as animal feed mills, transplant raisers and certain fertility and pest control inputs. This makes the system quite unique in that it prescribes standards for the whole of the food chain.

11.2.4 Product formulation

It is not just a case of substituting non-organic ingredients with organic ingredients. EU organic standards severely limit ingredients that can be used in organic processed products. Organic food products are currently divided into two main categories: 70–95% and 95–100% organic agricultural ingredient content. The former products cannot be described as organic in the sales description but ingredients can be identified as organic in the ingredient declaration. The latter products can be described as being organic in the sales description but must include a reference to the agricultural ingredient as obtained from the farm, for example organic yogurt made from organically produced milk.

Ingredients used in organic foods are divided into the following categories: organic agricultural, non-organic agricultural, non-agricultural and processing aids. Non-organic agricultural ingredients can only be derived from a limited list published in the standards. These include certain starches, vegetable oils (excluding sunflower and olive which must be organic) and a few other minor ingredients. Non-agricultural ingredients include salt and water and a limited range of food additives which excludes preservatives (except sodium nitrate and nitrite) and colourings. Certain processing aids are permitted but again this list is very limited. Full details can be obtained from the Soil Association Standards for Organic Food and Farming.

11.2.5 Process restrictions

Although it is preferred that organic processing takes place at dedicated organic processing facilities, it is acceptable for processing to be undertaken at non-dedicated facilities providing comprehensive separation and documentary procedures are applied. Bulk powder handling and continuous flow operations tend to offer the greatest challenges for integrating organic production into a non-dedicated process. Most standard food industry unit operations are permitted for organic foods but with processing becoming ever more technically diverse, there is an increasing pressure for certain processes to become restricted. Practices such as mechanical recovery of meat are prohibited but other processes are coming under increasing scrutiny such as standardisation and homogenisation of liquid milk.

11.2.6 Challenges for auditing

The greatest problem facing the inspector is that organic food looks the same as non-organic. Gone are the days of irregularly shaped, pest-damaged fruit and vegetables as lessons have been learned in production techniques and grading. This makes documentation and traceability the only method of identifying a product's organic status. Agro-chemical and veterinary residue analysis is impractical for verifying organic status due to the expense, the wide number of agro-chemicals used and the fact that produce may be subject to contamination during production, storage and distribution. Organic standards only prohibit the use of such materials but do not guarantee that the end product is free of these materials. The standards do have requirements for minimising contamination and operators are required to report any suspected contamination.

The range of processing operations is large ranging from small on-farm processing operations such as veg-box schemes to multi-national processors supplying major multiples. This equates to a wide range of operator ability as small operations tend not to have technically qualified staff. Therefore the inspection has to be 'pitched' at the level of operation being inspected. Other challenges for auditing include the relative lack of experience of operators in organic processing due to the rapid growth of the industry in a relatively short period.

Because demand is outstripping supply, there are elevated premiums for organic produce with commodity prices up to treble the price of non-organic produce which feeds through to retail sales. However these premiums are also necessary for producers, as organic agricultural production is a much more expensive method of production. Due to these increased premiums and shortage of organic raw commodities and ingredients, temptation to supplement organic ingredients with non-organic ones must be increasingly attractive both to meet customer demands and take economic advantage of the current premiums making the inspectors role increasingly challenging.

11.3 Certification and the auditing process

The authors only have direct knowledge of the procedures and policies of Soil Association Certification Ltd and although the content of the following sections may be applicable to other bodies this should not be assumed. Where the word 'Standards' is referred to within the text it specifically relates to the Soil Association Organic Standards for Food and Farming.

11.3.1 Certification

The complexity of introducing a new organic product into an operation will obviously vary considerably but as already stated it is rarely as straightforward as substituting organic ingredients for non-organic ones in established brands. There are therefore some key considerations to be included in the planning process (Table 11.1).

Time scales

It is crucial for operators to plan enough time into the development process to allow for certification and to ensure product development plans and launch dates are realistic

Choice of certification body

The operator will need to select a certification body authorised to undertake organic certification. There are currently 10 in the United Kingdom to choose from:

- Soil Association Certification Ltd
- UKROFS – United Kingdom Register of Organic Food Standards
- OF&G – Organic Farmers and Growers Ltd
- OFF – Organic Food Federation
- SOPA – Scottish Organic Producers Association
- Demeter (Biodynamic Agriculture Association)
- IOFGA – Irish Organic Farmers and Growers Association.
- Food Certification (Scotland) Ltd (Certification of farmed salmon only)
- Organic Trust Ltd
- CMi Certification (recently approved)

Table 11.1 How to gain organic accreditation

Accreditation step	Documentation	Notes
Choose certification body	Obtain copy of organic standards	Compare various standards and charges
Advisory visit		Optional but recommended
Prepare and submit application	• Product specification sheet(s) • GMO declarations • Supplier certificates, • labels or artwork • Cleaning schedules • Pest control details • HACCP	• Do market research • Source suppliers • Formulate product • Assess process
Receipt of application acknowledged and inspector allocated	None	
Inspection	• Quality manual • Training records • Goods in records • Production records • Sales/output records • Delivery notes. Invoices	• Report produced listing any non-compliances and recommended actions • Copy left at site
Report submitted for consideration by Certification Department		• Specifications checked • Certification equivalence of ingredients checked • Any other issues decided upon
Compliance form issued detailing actions	Send in any requested information	• Sign and return
Organic licence issued		• Both process and product(s) licensed. • Licence lasts for 12 months
	Organic accreditation achieved	

Of these, currently Soil Association Certification Ltd is the only UK body accredited by IFOAM. This means that products can be exported more easily as other countries would recognise the equivalence of IFOAM standards.

Some bodies have developed their own set of organic standards according to the aims and principles of the body, but in all cases meeting the requirements of the EU Regulations.

Preparation
The initial preparation prior to submitting an application is crucial if frustrating and costly delays are to be avoided. The more of the required information provided at the outset the easier the certification process will be.

Sources of information
Reading the standards and advisory packs issued by certification bodies may well provide sufficient information but consideration should also be given to advisory visits. Some certification bodies provide an advisory service which should be considered early in the planning process or alternatively an operation may wish to use the services of a recognised independent consultant. However, there is a limited range of these with appropriate experience due to the relative 'youth' of the industry.

It should be noted that the approval of a product is complete only when the operator has been issued with a Certificate of Registration, which allows for the production of only those products listed at the registered site shown. The premature sale or marketing of non-licensed organic products is a legal offence.

11.3.2 The auditing process

The actual process of the audit may vary slightly according to the preference of the inspector but should be consistent with the outline described here. A more detailed description of what the inspector will be looking for is included in the following section.

Confirmation letter
Once the date and time of the inspection has been agreed the inspector will confirm these details in writing. The inspector will also include details of the information that will be required to allow for the completion of a comprehensive audit. The certification body will supply the inspector with a copy of the application form or previous report and supporting information depending on the nature of the audit.

Introduction
On arrival at the site the inspector will briefly explain what is required and agree a logical programme for the audit. This will ensure that, where applicable, a physical inspection of the process can coincide with a production run, the appropriate staff will be available when required and any documentation requested can be compiled for review during the audit.

Confirmation of current status
A check of the product range including (where applicable): product specifications, labelling, status of suppliers, import authorisations and GM declarations. A review of documented procedures and training.

Physical inspection of premises
A tour of the site which will include all stages of the process from 'goods in' through to despatch and a focus on those parts of the operation where there could be a risk of loss of the product's integrity. The storage of cleaning chemicals and pest control materials will also be noted.

Documentation
A detailed check of input, production, output, cleaning and pest control records to confirm the information supplied verbally and that documented procedures are being adhered to.

Audit trail and mass balance analysis
At least one example of a product will be taken, either physically from storage or from the records, and traced through the process records to ensure that integrity has been maintained. A check will also be made to ensure that quantities of ingredients used correlate with the amount of finished product produced from them.

Report and summary
The inspector will produce a report on site and summarise the findings. Any compliance issues will be raised and the actions arising explained. It should be noted that the inspector's role is to gather information and make recommendations. They are not allowed to give advice and they do not make the final decision about whether the licence should be awarded.

Types of audit
As well as initial and annual audits the certification bodies have a requirement to complete a number of unannounced audits and may conduct additional arranged audits to confirm that recommendations made by the certification body have been implemented.

UKROFS also have a legal entitlement to carry out unannounced surveillance audits of licensed operators but the purpose of these is monitor the performance of the certification body inspector rather than the production process.

11.4 What auditors look for (on site at an inspection)

11.4.1 Purpose of inspection

Organic certification relies on inspection to ensure that standards are being complied with. The purpose of inspection is to gather sufficient information to make an appropriate decision on licence award/renewal. It also enables a reaffirmation of company culture and the licensee's attitude to compliance with the standards. Auditing is a tool that is used by the inspector as part of the inspection to 'test' aspects of the system to ensure that standards are being complied with consistently to a satisfactory level. The organic standards require that 'the inspector must make a full physical inspection, at least once a year, of the organic unit'. During the inspection the inspector has to satisfy many demands:

- Statutory and organic certification requirements.
- Requirements of the certification body which are additional to the basic standards, for example correct use of logos and verifying level of sales values for fee assessment.
- A good inspection is also viewed as one where there is a two way flow of information however inspectors are prevented from 'giving advice' other than making recommendations for improvements to comply with standards.
- The inspector is also indirectly there for the benefit of the consumer, ensuring that when a consumer picks a product off the shelf, they can be assured it has been produced to the organic standards.

11.4.2 Dedicated and non-dedicated organic processors

The standards recommend organic processing operations be dedicated to organic processing. However desirable, this is often not practical in a transitionary phase while markets are developing and new entrants are coming into the industry.

From an inspection point of view, it is not whether an operation is dedicated or not that bears on its ability to maintain the integrity of organic products but how well it is managed. The fact whether an operation is dedicated or not has a slight influence on inspection procedure but the emphasis on the key points does not change.

It is non-dedicated organic processors who form the bulk of organic registered processors. In order to maintain the integrity of organic products in terms of cross contamination, it is essential to ensure that there are full separation procedures in place from raw material receipt to finished product dispatch. An assessment of the necessary separation requirements is undertaken at the application inspection. This will relate to storage areas, preparation, processing, packing and final storage areas. Recommendations are made to ensure that full separation and identification will remain throughout. Application of these recommendations can be audited objectively by analysis of production records, physical inspection of storage facilities and a check on variance of organic ingredients purchased and used.

Physical inspection of storage facilities would concentrate on clearly separated and labelled areas which should be dedicated to organic product. Examples of storage separation would be:

- flour silos to be totally dedicated to organic product due to product clinging and difficulty of cleaning;
- milk silos can be dedicated by time and cleandown as they are easy to clean in place (CIP).

Generally separation is achieved in the processing areas by processing as first operation after a full cleandown. This is checked at inspection by inspecting production records to ensure cleaning has been done prior to organic production or more preferably, that there is a sign off procedure by responsible staff prior to organic production. Certain operations cannot be effectively cleaned by CIP or manual cleaning so bleed runs will be required. Examples of this are dry bulk powder processes, chocolate machinery and items such as fat pumps for pastry production. Generally to verify that this is being done, production records will be analysed and quantities of organic product used are compared with quantities of product made. The quantity required for a bleed run will need to be agreed with the certification body and is likely to depend on the size of the plant and type of product.

Separation is also enhanced by requirements for certain utensils to be dedicated, particularly absorbent materials such as plastics in contact with fatty foods. Fortunately, with the large range of coloured plastics available, colour identification is simply achieved and colour coding is used extensively in the organic industry. Not surprisingly, green seems to be the preferred colour.

11.4.3 Preparation for inspection

If standards are being applied competently, management is effective and operators are given adequate training and are sufficiently motivated, then preparation for inspection should involve a minimum effort. However, some prior preparation will make the inspection flow more smoothly for both the licensee and the inspector. Inspectors try to give 3–4 weeks notice, initially by phone, followed up by a letter confirming the appointment and giving details of what will need to be seen at the inspection.

The licensee can best prepare by following the points below:

- Ensure relevant departments and personnel are aware of the inspection and can be present on the day.
- Ensure that records relating to all purchases are available. If they are held at a head office, they may need to be brought to the site for the day.
- Records relating to all organic production runs should be available. Using colour coded paper for organic records helps in identifying them if they are filed with the main records.
- Ensure other relevant documents are available such as work instructions, procedures/quality manual, training records, GMO declarations, specifications of additives/processing aids, examples of all labelling.
- Records relating to hygiene and pest control since the last inspection need to be available.
- Summaries should be produced of total quantities of organic ingredients purchased, used and sold for the period since the last inspection or for the company financial year.
- Hygiene and health and safety clothing must be made available.

11.4.4 Organic ingredient verification

This is a key part of organic operation and is often overlooked by processors. Companies often pay a much higher price for organic ingredients and for this reason alone it would make sense to do some basic checks on the product to ensure it is what they have paid for when it arrives on site. The certification body checks organic certification of ingredients when a product specification is submitted for approval. The inspector will check the following at inspection:

- The supplier and certification of ingredients being used are the same as those detailed on the specification sheet.
- Delivery documentation/purchase invoices clearly state the product is organic.
- Packaging and labelling clearly indicates the product is organic.
- Imported product from outside of the EU is from a correctly licensed importer and is accompanied by EU import certificates.

Purchase documentation and stored product labelling will be checked to ensure that they clearly state raw materials are organic as this will be the only proof that organic ingredients have been used. Goods in records will also be checked to ensure that the result of organic verification check is explicitly mentioned in the goods in records or equivalent. It is considered best practice for licensees to obtain and keep on file, a copy of the up to date organic certificate for the product and supplier. Certification of products and suppliers has to be differentiated because wholesalers who do not repack product do not need to be certified, so ingredients purchased from these suppliers need to have their certification verified back to point of last operation. It must not be assumed that the intermediate wholesaler has checked the certification as they are under no obligation to do this. Companies need to be aware that the demand for organic ingredients and high dependence on imported product adds to the risk of fraud and close monitoring of the status of imported ingredients is therefore essential.

11.4.5 Mass balance and traceability audits

The two key exercises that are undertaken at annual inspection are the traceability audit and a mass balance or variance audit.

Most companies are used to undertaking internal traceability checks and all food companies are legally required to have some sort of system in place to ensure that ingredients in a finished product can be traced back to source via coding of ingredients. The true test of a traceability system is that ingredients can either be traced back to source from a product code, or can be traced forward to a product via an ingredient code.

One tool that is used world wide to check the organic integrity is the mass balance or variance audit. This check is done to ensure that sufficient quantity of organic ingredient is purchased to make the quantity of product that has been sold. Annual summaries of quantities of organic ingredients purchased and product made should be kept to assist this check. Opening and closing stocks of ingredients and products are also taken into consideration. Most computer systems enable this sort of information to be generated easily but manual records need careful planning to ensure this information can be easily generated. Product specification sheets provide information of quantities of ingredient used for a given amount of finished product to enable the inspector to make the calculations.

Obviously, it would be rare for ingredients purchased to exactly match product made as there are various production losses, however providing there is marginally more organic ingredient purchased than required for product made, then this will be satisfactory. If there is a larger difference than might be expected then this would need to be explained. Detailed records of production losses, trials and samples should be kept.

If insufficient organic ingredients have been purchased compared to the quantity of product made, then this is more serious as it could indicate that perhaps organic ingredients have been substituted or supplemented by non-organic ingredients. This could either be accidentally or deliberately. If it is discovered that there is a shortfall of organic ingredient and it cannot be explained satisfactorily, then this may result in a particular product or production run having its organic status removed. Depending on the seriousness of the problem, it could result in the whole licence being suspended or revoked. If a company is dedicated to organic production, then this is effectively the termination of the business activities unless a non-organic market for its products can be developed rapidly.

11.4.6 Product labelling

Product labelling will be checked to ensure that it complies with regulatory requirements and the individual certification body standards and that the ingredient declarations correspond with specifications held on file. Checks will also be made to ensure that examples held by the certification body match those in use by the licensee. Examples of all labelling should be available for inspection.

11.4.7 Cleaning, hygiene and compliance with statutory requirements

Although this is primarily checked by Environmental Health, Trading Standards and various other audits, the inspector will note any areas of hygiene that are not considered appropriate for processing organic foods as requiring a compliance action. It is a requirement of the Standards that all other relevant food legislation is met as a minimum, such as premises registration and food safety legislation.

Cleaning records are checked to ensure that appropriate cleaning has been done both before and after organic processing. It is a requirement that terminal sanitisers are rinsed off prior to organic production. Cleaning records and sign off procedures will need to demonstrate this has been done.

11.4.8 Pest control

Pest control records will be inspected to ensure that no prohibited materials have been used and that any materials that are used have been used responsibly. Rodents get no reprieve from organic standards but insect control materials are severely limited. Preference is given to good housekeeping and physical exclusion measures. Pest records are also checked to ensure that any proofing/housekeeping actions recommended by pest contractors have been actioned effectively. Again, signs of good housekeeping and pest activity will be looked for during the tour of the premises. Where contamination is suspected, then samples may be taken for analysis. Use of prohibited materials, pest contamination of product or inadequate recording of pest control materials used could result in either production runs having their organic status removed or the whole licence being suspended.

11.4.9 Knowledge of standards

One area that is assessed is understanding and knowledge of the standards. It is a requirement that staff are given training to ensure they are competent to carry out their assigned tasks and understand the importance of maintaining the organic integrity of the raw materials and the finished products throughout the production cycle. This is assessed by level of compliance with the standards in general and by talking to production operatives to assess their knowledge of organic operating procedures. It is not necessary for operatives to know standards word for word, but a general understanding of the aim of the standards and how they can be applied in the context of their area of responsibility would be expected. It is no use having written procedures if the people who are actually doing the job do not know or fully understand what they are supposed to be doing and have not had its purpose effectively explained.

11.4.10 Attitude and competency of management and staff

The inspection is not just about comparing the physical operation against standards and making various recommendations, it is also about trying to evaluate the attitude of the management and staff. The company culture and attitude of management and staff is often the first impression that is gained by the inspector upon arrival at site.

General tidiness of the site and quality of preparation prior to the inspection gives the inspector an initial impression of the company culture and likely compliance with standards.

If the attitude is not considered appropriate, this is addressed by encouragement and dialogue and by creating good working relationships; however, the relationship should not become too familiar. Where non-compliances are found, the seriousness of the transgression and any relevant penalties will be made clear.

Competence is checked by subjective and objective assessment of general implementation of standards, record keeping, preparedness for inspection and correction of non-compliances from previous organic inspections. Competency of management is a key factor as this gives indication as to how well standards are applied during the year when the inspector is not there. It was suggested by Bowles (2000) however, that the attitude of management has a greater overall influence on compliance than competence. Where the inspector suspects problems, follow-up inspections or even unannounced inspections may be recommended to ensure that standards are being applied throughout the year. Where it becomes obvious that organic integrity of products may be at risk, then increasing penalties will be imposed eventually leading to a suspension of the licence until such time that management can demonstrate compliance with standards. If they are unable to do this, then the licence will be terminated.

11.4.11 Adherence to principles

This can be assessed subjectively by the overall view of the company's approach to organic policy. This would manifest itself in type of product made, formulation, where ingredients are sourced from, level of dedication of processing plant, staff training, efficiency of process and disposal of waste products. Although certification award/renewal is entirely based on compliance with standards, adherence to organic principles will be encouraged and may become an increasingly important factor in the future as the market and ingredient supply chains become better developed. Many organic consumers place great emphasis on this when making their purchasing decisions, they are after all, often paying a premium because they believe the product has been produced according to a set of principles.

11.5 Summary and future trends

The organic processing sector has coped with dramatic growth in recent years and consequently the certification process has had to keep pace with this change and adapt accordingly. This has resulted in improved provision of certification services and the appointment of inspectors with specific experience of food processing. Further developments are likely and these may include:

- A consolidation of certification bodies resulting in a more uniform implementation of the regulations.
- Greater use of information technology allowing full traceability back to the primary producer. Use of 'genetic fingerprinting' by DNA techniques could be used for traceability purposes or developed as a test for organic integrity.
- Self assessment becoming a part of the annual audit.
- An increase in the number of unannounced audits.
- A greater use of pesticide and antibiotic residue testing where presence is suspected.

- More emphasis on the equivalence of production methods used for imported organic products.

Organic production therefore involves considerably more than simple substitution of raw materials and a careful consideration of all the issues raised by this chapter is required before introducing an organic product and applying for a certificate of registration. Having made the decision to go ahead however food processors may have confidence that the requirements of the regulations need not be impractical to implement and are overseen by bodies and individuals with a good understanding of their needs and processes.

11.6 References

ANON. (1999) *The Organic Food and Farming Report 1999*, Soil Association, Bristol, p. 27.

ANON. (2000) *IFOAM Basic Standards*, Basel, Switzerland.

BOWLES, A. (2000) Presentation at IFST Annual Conference, NEC 04/10/2000.

TAYLOR NELSON SOFRES (2001) 'Rewriting the Record Books', *Organic Focus* Issue 16, 8, Reed Business Information, Sutton, Surrey.

Index